COMMON GROUND

the sharing of land and landscapes for sustainability

Mark Everard

Zed Books

LONDON | NEW YORK

Common Ground: the sharing of land and landscapes for sustainability
was first published in 2011 by Zed Books Ltd, 7 Cynthia Street, London
N1 9JF, UK and Room 400, 175 Fifth Avenue, New York, NY 10010, USA

www.zedbooks.co.uk

Set in OurType Arnhem and Futura Bold by Ewan Smith, London
Index: ed.emery@thefreeuniversity.net
Cover designed by Lydia Fee www.lydiafee.co.uk
Printed and bound in Great Britain by the MPG Books Group, Bodmin
and King's Lynn

Mixed Sources
Product group from well-managed
forests and other controlled sources
www.fsc.org Cert no. SA-COC-1565
© 1996 Forest Stewardship Council

Distributed in the USA exclusively by Palgrave Macmillan, a division of
St Martin's Press, LLC, 175 Fifth Avenue, New York, NY 10010, USA

A catalogue record for this book is available from the British Library
Library of Congress Cataloging in Publication Data available

ISBN   978 1 84813 963 3 hb
ISBN   978 1 84813 962 6 pb

Contents

Figures, tables and box

Acknowledgements

The author offers his particular thanks to Dr Tom Appleby (UWE), Dr John Colvin (Open University), Sam Chimbuya (Khanya-aicdd), Bill Watts (Environment Agency), Dr Mervyn Bramley, Ken Tatem (Environment Agency), Janina Gray (Salmon and Trout Association), Dr Paul Raven (Environment Agency) and Dr Debbie Pain (Wildfowl and Wetlands Trust) for their contributions to the streams of thinking, many leading to co-authored scientific papers, underpinning this volume.

The author would also like to acknowledge the staff of DWAF KZN (the Department of Water Affairs and Forestry, KwaZulu-Natal provincial office) participating in the UK government-sponsored 'Watercourse' development programme, and the citizens of the Mvoti catchment (KZN) who engaged in an action research process aimed at developing practical, integrated approaches to catchment management in KZN.[1] Thanks are also due to other colleagues with whom the author has worked in the practical capacity-building elements of the work in South Africa featured in this book, including Faeeza Ballim (IWR Water Resources), John Goss (Cinnabar Global), Geraldene Klarenberg (Mvula Trust), Sizile Ndlovu (formerly of the Inkomati Catchment Management Agency), Dumisani Ncala (Cinnabar Global) and Derek Weston (Pegasys). Relevant work undertaken in India benefited greatly from collaboration with Gaurav Kataria (www.india-angling.com) and Steve Lockett, while work in China has been in collaboration with Professor David Lerner and Professor Bob Harris (both of the University of Sheffield).

Thanks too to Jakob Horstmann and Tamsine O'Riordan at Zed Books for their encouragement and helpful comments.

The efforts of all of these people, and the processes that we have co-created, have been a primary impetus in the development of this book and the canon of thinking underpinning it. All of this creativity and application brings us together within the common mission of accelerating progress to a just and enduring future for all of humanity and the landscapes that support us.

Foreword

Mark Everard has one simple message for us in *Common Ground*: 'It's all about the land, stupid!'

Or rather, it's all about the connectivity between the land, water, climate, energy, culture, economy and human creativity. So don't go sticking this one into the pigeonhole marked 'Environmental Problems' – because that isn't what you'll unearth in *Common Ground*.

What you will find are timely reflections on a world transitioning between different paradigms – from a world in which reductionist, issue-specific policies (that led, inevitably, to all sorts of so-called 'unanticipated' consequences) dominated, to a world in which much more sophisticated systems thinking is driving integrated policy-making and implementation.

It's an uncomfortable transition. Even as our natural scientists gain confidence in more integrated thinking, conventional economic policy is still fiercely and unapologetically reductionist – bringing dangerously redundant methodologies (such as orthodox cost/benefit analysis) to bear on complex, multifaceted land-use dilemmas. As Mark Everard says: 'Our economy and its vested interests remain stubbornly and reductively divorced from the ecosystems that ultimately support it.'

As a result, our market-based economies simply aren't delivering. Many of the environmental services that society has at last begun to appreciate more realistically (carbon management, climate regulation, biodiversity, water and so on) go entirely unrewarded by conventional markets. The idea of 'payment for ecosystem services' has come a long way, conceptually, since the publication of the Millennium Ecosystem Assessment (MA) in 2005, but politicians and economists are still struggling to convert this wealth of scientific analysis into practical land and resource management tools. Six years on, one still has to ask whether the MA's fundamental warning has in fact been heeded: 'In the midst of this unprecedented period of spending the Earth's natural bounty, it is time to check the accounts. That is what this Assessment has done, and it is a sobering statement, with much more red than black on the balance sheet.'

The logic that lies behind the imperative of treating ecosystem services as critical natural capital is unanswerable. The direct and indirect economic benefits we derive from those services depend entirely on maintaining the capital assets intact. Mark Everard offers us a range of case studies (drawn particularly from the UK and South Africa) which show how this logic can be translated into action. But the truth of it is that dominant economic mindsets today are changing more slowly than the accelerating environmental damage done by more and more people as they seek to get richer and richer. There are often complex political reasons why that remains the case:

> The biophysical reality that humanity is fully dependent upon the services provided by ecosystems may have become buried in societal consciousness under assumptions about private land rights, the annexation of crucial resources such as soil, water, minerals and biodiversity by a privileged minority, market pressures to liquidate common assets for short-term gain, and the global reach of empire-building and its analogue in powerful multinational businesses.

This raises the question of the kind of governance systems that will need to be put in place to promote improved decision-making on land-use issues. Politicians tend to give up their powers very reluctantly, and are still much more inclined to pursue sham consultations rather than find ways of genuinely involving key stakeholders in complex land-use decisions. DAD (decide-announce-defend) remains the order of the day, with only a few inspiring models of EDD (engage-deliberate-decide) available to us to demonstrate how very different things could be.

To many, this will still sound very utilitarian and 'processy' – especially for those activists who remain deeply sceptical about the whole notion of 'marketizing' the natural world in order to save it. Mark acknowledges those tensions, drawing on the land ethic of Aldo Leopold (which '... enlarges the boundaries of the community to include soils, waters, plants, and animals, or collectively: the land') and the 'deep ecology' of Arne Naess to remind us that the cultural and spiritual 'services' that ecosystems furnish us with are in many ways as important as the biophysical and economic services.

> Many aspects of our cultural journey have separated us from a mutually supportive relationship with the land and the ecosystems that provide for our needs. Incrementally, Western science and philosophy are beginning to acknowledge that the dependence between humanity and ecosystems with which we co-evolved is not only central to our basic biological needs but, as many other cultures around the world accept as axiomatic, vital to our mental and spiritual well-being.

We should celebrate the fact that these insights sit quite comfortably alongside today's unprecedented wealth of evidence-based insights from a new generation of biologists, agronomists and hydrologists – all seeking to make sense of more integrated, sustainable land-use practice. And that's exactly what *Common Ground* sets out to do.

Jonathon Porritt
Founder Director, Forum for the Future
21 December 2010

Introduction

The landscapes that humanity has inhabited throughout our evolutionary history have simultaneously met our needs and substantially shaped
our progress. We too, through our gifts of innovation and foresight
(as indeed our lack of it), have substantially shaped the landscape,
atmosphere and waters of this resilient yet finite world, appropriating
increasing proportions of the resources of nature to meet our shifting
and growing demands. And thereby we have progressively, if largely
unintentionally, undermined the quality and extent of the ecosystems
that support us, including their various capacities to meet the needs
of our burgeoning population.

Our relationship with the landscapes we inhabit is central to our
past, present and future. This book refers frequently to the terms 'land'
and 'landscape', for each of which there exist many definitions. For the
current purpose, the term 'land' is taken to refer to the solid part of
the Earth's surface, including its many attributes and functions, such
as soil, hydrological properties and places for human habitation and
food production. The definition applied to the word 'landscape' draws
from its meaning within the term 'landscape ecology' initially coined
in 1939[1] to describe interactions between environment and vegetation
but subsequently broadened to address heterogeneity within landscapes,
including spatial structure, distribution and abundance of organisms,
and the functions that it performs,[2] including the geomorphological
processes that shape it[3] and associated uses by and pressures from
humanity.[4] Neither term refers to ownership of land, but both are intimately associated with supporting not only diverse ecosystems but
also many societal interests. 'Landscape' thus refers to an arrange-
ment of heterogeneous habitats and the functions that this complex
socio-ecological system performs, while 'land' refers principally to the
biophysical resource itself.

The relationship between humanity and the critical resource of land,
and the integrated functions of diverse land types across the wider landscapes that support us, has shifted continuously throughout our tenure
on Earth, brief as it has been relative to the evolution of planetary life.
Notwithstanding our total dependence on the Earth's natural resources,

industrial-age humanity and the pervasion of industrial and capitalist ideologies throughout increasing proportions of the Earth's human population have already overridden many of the supportive limits of this world. They are now doing so at an increasing pace, with adverse and increasingly well-understood consequences for all who share it and depend upon it now and into the future.

This book sets out to explore the shifting relationship between human society and the resource of land and the wider context of the landscapes that shape and bear us. The first chapter, 'The privatization of the land', addresses the dependence of early civilization on common landscapes and their progressive annexation into private property throughout much of the world, with an associated loss of commoners' rights. Chapter 2, 'Reclaiming the common good', then traces the progressive recognition of the many public benefits flowing from land, regardless of who owns it, and the trace of legislation, common law and subsidies that are beginning to 'reclaim' these public goods, using primarily European and American examples.

Chapter 3, 'The ends of the Earth', refers to the ways in which exploitation of ecosystems beyond the limits of their carrying capacity contributes to many of the pressing sustainability challenges that we face today. This raises troublesome long-term implications for all who share the global 'common'. Chapter 4, 'Shifting conceptual landscapes', then turns to transitions in human understanding and exploitation of the natural world, and how these have shaped perception and response to development, and latterly sustainable development challenges, primarily throughout our industrial past but looking ahead to what will be required in future.

Chapter 5, 'A landscape at our service', introduces the evolution of 'ecosystem services'. Ecosystem services define the many benefits that people derive from ecosystems, offering insights into the dependence of all human interests on the natural world and making a substantial potential contribution to finding a more sustainable relationship between society and the landscapes that support it. Chapter 6, 'The great food challenge', then explores our changing relationship with land from the perspective of food supply. Chapter 7, 'Valuing land and landscapes', considers valuation of ecosystem resources and how they can be used to develop more inclusive markets to support sustainable interactions with the land that supports us.

This is then synthesized in Chapter 8, 'Living landscapes', to consider management, markets and other tools and exemplars of best practice from around the world that can accelerate progress towards sustainability through the integration of different perceptions and value systems, leading to the creation of economically, ecologically and socially connected 'living landscapes'. Chapter 9, 'Lessons for tomorrow's world', distils principal lessons from the preceding chapters to guide future development with respect to our relationship with land, with Chapter 10, 'The people's land', then concluding this book with moral and cultural considerations relating to the achievement of sustainability.

Although the ultimate destination of this book is to consider what is necessary to secure a sustainable relationship between humanity and the landscapes that support our needs on an enduring basis, we start off by looking backwards to understand more of this ever-shifting relationship.

1 | The privatization of the land

The roots of humanity are well obscured by time. However, we can be certain that the dependence of early civilizations on landscapes will have been a defining feature. The best evidence that we have suggests that much of our prehistory comprised nomadic or semi-nomadic hunter-gatherer communities that moved opportunistically with the availability of food and the avoidance of natural hazards, probably lacking any sense of land ownership. Of necessity, these communities would have been small and, generally, closely related groups.

We may glean some insights about this phase of our past from the few nomadic communities still to be found outside of the dominant economic model that defines much of the modern world. We may then proceed to consider subsequent developments of society, including those arising from the settlement of communities and the ownership of land.

Mobile communities and the land

There are thought to be between 30 and 40 million nomads in the world today.[1] Although truly nomadic people are rare in Europe, they are to be found across much of the world. In Africa, the San (or Bushmen) of southern Africa are just one example of the diverse nomadic communities scattered across the continent. Australia hosts some aboriginal communities still not significantly influenced by Western culture. In Asia, Iran is home to one of the largest nomadic populations in the world with an estimated 1.5 million within a country with a total population of about seventy million,[2] and there are widespread other nomadic communities across the continent, including, for example, the Adivasi tribal people of India, who form part of the estimated 0.7 per cent of the Indian population who live nomadic lifestyles.[3] In the Americas, there remain 'First Nations' tribes in both North and South America who still cleave to traditional nomadic lifestyles, albeit that many in the USA and Canada do so largely within prescribed reservations.

It is far from clear how early nomadic communities viewed ownership of the land that they inhabited. Famously, American First Nation 'Indians' are often quoted as proclaiming themselves merely short-term stewards of a timeless landscape. Perhaps the best-known among these quotes is the 1854 response by Chief Seattle to a government offer to purchase the remaining Salish lands,[4] relating to their spiritual connections and symbiotic relationship with this land. Zulu and many other tribal discourses speak of the people belonging to the land, which ancestral spirits also inhabit, rather than of the land belonging to people. Nomadic people from around the world would respond to changing seasons and other conditions, moving across landscapes that were not owned by any person or tribe. However, they must have had protocols for planned or accidental encounters with neighbouring tribes, which represented not only competition for scarce resources but also companionship and possibly an exchange of mates.

Establishment of common rights

There is no single accepted definition of what exactly constitutes 'common land'. However, the term is generally used to describe a piece of land in either state or private ownership to which other people have certain traditional rights to use it in specified ways, such as being allowed to graze their livestock or gather firewood.[5] This differs from 'communal land', which generally refers to (mainly rural) spaces in the collective possession of a community, rather than an individual or company. Those who possess rights to use a common are known as 'commoners', although landowners may retain other rights to the land, such as rights to exploitation of minerals and large timber, and any other common rights left unexercised by the commoners.[6]

Globally, cooperation is a prevalent feature of tenant communities living on land shared as a common property, or owned collectively as 'communal land'. Today, for example, one can still venture into more remote reaches of the Western Ghats, towards the western seaboard of peninsular India, and observe communally managed, centuries- or millennia-old terraced paddy landscapes in forest river valleys, where ownership of the land itself is either indeterminate, vested in the state or, more commonly, in the hands of absentee private landlords. It may not be safe to assume that early human communities observed this same structured means for the exploitation of land, but it is certain that some degree of cooperation would have been necessary to avoid conflicts between users.

Settlement of communities and changing relationships with land and landscapes

The foundations of what we tend to consider today as modern civilization were laid by the domestication of crop plants and stock animals.

Artefacts of irrigation and river management practices found widely across the ancient human world provide evidence of one of civilization's founding technologies: the manipulation of water flows to and from land to enhance natural soil productivity and overcome limitations to human survival and progress. Settled agriculture, in turn, marked a revolution in human ingenuity. It signalled an evolution in human consciousness to the extent that farmers' decisions were premised upon future yields rather than merely current availability. It created the need for the establishment of permanent communities, with new levels of social structure and communication, and for the transfer of knowledge between individuals and generations. These innovations laid some of the foundations of modern culture. Writing systems were developed, with many early examples applied for the purpose of accounting for stores of agricultural produce.[7] The origins of economic systems are generally ascribed to the transition to settled agriculture as, released from the daily drudgery of pursuing food, societies were able to differentiate responsibilities and labour and so to require a trading system for fair exchange between social groups.

From these settled communities, empires were created, often in the quest for more resources to support booming populations and denser towns sustained by stable agricultural productivity. This in turn enabled the building of civilizations by better-nourished and organized people. Monumental buildings were constructed by an adequately fed populace less vulnerable to the vagaries of hunting and gathering, and with a greater luxury of time to devote to symbols of status and religion.

Changing land use and rights in Europe

Use of and rights to land are best documented in Europe, and so this will be the focus of the consideration of communal rights that follows. Although Europe has a long history of settlement, invasion and experimentation with different governance models, much of the landscape nevertheless remained effectively in common stewardship in the era of low population preceding the 'early modern period' (roughly the start of the seventeenth century).

In the absence of strongly centralized and hierarchical social systems, large swathes of European land were managed collaboratively as a common resource, generally by village communities.[8] In England, collective farming of common land was once a widespread practice, probably covering more than half of the farmed land area of England in medieval times. Contrary to the views expressed by some historians and economists, for example in Hardin's metaphor of the 'tragedy of the commons',[9] which implied that land not owned by its users was likely to be subject to progressive over-exploitation and degradation (we will return to the 'tragedy of the commons' later in this book), stewardship systems on common land were often highly structured across medieval and early-modern western Europe (from the sixteenth to the nineteenth centuries).[10] Conflicts arising from the use of land, water and other natural resources were largely resolved at a local level. Manorial court records provide a rich source of case law relating to resource conflicts, particularly with respect to access to and use of water resources from the post-medieval period as the spread of the once-dominant water meadow systems in the catchments of Wessex rivers spurred conflicts between water meadow operators, mill owners, navigation and fishery interests and other users of river flows.[11]

This was backed up by statutory regulation that supported both private rights and the wider public benefits of common resource use. The earliest recorded water quality regulation dates back to 1388 in the shape of a royal statute declared by King Richard II prohibiting the dumping of 'dung, offal, entrails and other ordure into ditches, rivers, waters ...'. Legislation tended to operate both through local courts and evolving bodies of local customary law governing rights such as grazing and hay-cutting. Often, these were formalized into village or local by-laws which, in turn, served to inform decisions achieved by the courts. This evolving body of local customary law developed at grassroots level, adapting to local conditions and thereby providing a legal framework of common rights responsive to local environmental, social and economic pressures. Similarities between sets of village by-laws over large parts of England suggests that they were constructed from the foundations of wider custom and culture,[12] and also adapted to broadly similar agricultural methods and supportive landscapes. A similar commonality of principles has been found to occur in long-enduring common resource management systems across the world.[13]

During this phase of English history, mirrored across much of western Europe (and we will explore other regions subsequently in this chapter), most families could claim some rights of use to common land. These could span any or all of a number of entitlements, including rights to access parcels of land on the arable acreage or pasture of the manor farm, gather hay or wood or graze cattle, or to hunt and fish on the 'wastes' beyond the common fields. Overall, this had an equalizing effect on English and European society.

Enclosure, possession and shifting societal rights

This relatively egalitarian access to land and natural resources became progressively overturned throughout western Europe from the thirteenth century onwards through to the nineteenth century under the process of 'enclosure'. The history of this is most comprehensively documented in England, which provides the example below.

Enclosure (though the more ancient spelling 'inclosure' still remains in some UK legal documents and place names) saw the progressive curtailment of various traditional rights of access to land owned by another person or held in common by numbers of people. Under enclosure, land claimed as private property became fenced (hence 'enclosed') or else was deeded or entitled to one or more owners. This process was instigated by successive Statutes of the Realm. The first and most significant of these was the Statute of Merton, agreed in 1235 at Merton between the barons of England and Henry III, which permitted the Lords of the Manor to '... make enclosures on the waste ...' on the condition that sufficient pasture remained for their tenants. The Statute of Merton was significant for various reasons, including establishment of the legal concept of 'ownership', which quickly formed a basis for English common law.

Enclosure often initially took the form of the piecemeal fencing or hedging of strips of land from the margins of common fields throughout the medieval and modern periods. Sometimes, this was undertaken by smaller landowners, but the majority of land take was executed by larger landowners, significantly including the powerful lords of the manor, for example to establish deer parks or other hunting reserves as well as to claim fertile farmland or else more marginal land which could be managed intensively without the benefits of conversion being enjoyed by former common claimants.

The history of enclosure was often bloody, with land sometimes taken by force in the face of strong local resistance from people effectively disenfranchised by exclusions. This frequently involved significant bloodshed. It also had radical social consequences, installing the class system of lords, vassals and fiefs that characterized the feudal system and which also, effectively, created a new landless working class.[14] Successive English parliamentary Acts of Enclosure from around 1750 to 1844 created stronger legal sanctions for the privatization of common rights. The rapid expansion of enclosure of agricultural land between 1760 and 1820 saw an accelerated loss of common rights, with a mass migration of this disenfranchised peasant underclass into burgeoning urban centres to meet the ever-growing labour requirements of the new factories of the industrial age,[15] particularly in the booming industrial heartland in the north of England. By the twentieth century, unenclosed commons had become largely restricted to rough pasture in mountainous areas and in relatively small parts of the lowlands. Notwithstanding other profound social changes throughout this period it is, on reflection, hard now to regard the age of enclosure as anything other than class robbery.[16]

This simple telling of a well-established story oversimplifies the uneven distribution of rights prior to enclosure, including a successive loss of land rights that may be traced back to the Norman or even the Saxon period. For example, William the Conqueror claimed one third of the land for his hunting estates after the Norman invasion of England.[17] In fact, 45–50 per cent of the agricultural land of England may already have been enclosed by 1500 and 70 per cent by 1700, substantially prior to the demands of industry and its own profound effect on the changing face of British society.[18] Indeed, some better-off members of the European peasantry appeared to have actively encouraged and participated in the process of enclosure, perhaps to help bring to an end the perpetual poverty of subsistence farming; enclosure may have been ascribed too much importance relative to wider societal change.[19] Enclosure is therefore best regarded not as a sudden form of oppression and appropriation of land rights, but rather as part of a sustained pattern of economic and other pressures associated with agricultural 'improvement' and social reform.

Enclosure of fens, marshes and other formerly agriculturally marginal land was to result in a net increase in total agricultural output. Yet whether an imposition by the wealthy or with complicity from some of the many rural poor, the loss of hunting, fishing and subsistence farming rights certainly was a matter of life and death for many western European people throughout the eighteenth and nineteenth centuries. For those who lost access to vital resources, hardships were real and formed part of radical changes in livelihoods, including urbanization and the creation of a significant powerless class of labourers available to rich agricultural or industrial entrepreneurs.

Enclosure was far from a uniquely British phenomenon. It was one that permeated much of western Europe. For example, Ireland was to become progressively enclosed from shortly after agreement of the Statute of Merton, and Germany was to follow much later in the nineteenth century with a process of 'Landflucht' (literally 'flight from the land'), which saw a mass exodus of rural workers into cities and their booming industrial enterprises. The socialist Friedrich Engels[20] was to compare the German Landflucht with English historical enclosures, drawing similar conclusions about the disenfranchisement of agricultural workers and the creation of a servile working class. Elsewhere in the world, control of key resources through the enclosure of land by ruling or influential classes was a frequent phenomenon, repeated from European colonial occupation of eastern Africa, India and beyond to apartheid-era South Africa. Even such apparently constructive practices as the establishment of national parks has been associated with significant concerns about 'environmental racism' in both South Africa[21] and the USA[22] and also in India, where the antecedents of modern conservation policy today can be found in the practices of princely India, with their hunting reserves, on occasion stocked with imported wildlife.[23] Indeed, there are widespread concerns that conservation policies marginalize or disempower local groups.[24] Ongoing 'land claims' by bodies and individuals from nineteenth-century US mining claims to modern-day Zimbabwe demonstrate the contentious history of enclosure.

Consolidating the rights of the oligarchy

Once the feudal ownership arrangements of UK land had been simplified through enclosures, privileged use and development of land by its owners predominated in the British Isles right through to the mid-twentieth century. For much of this phase of history in England, the metaphor of 'An Englishman's home is his castle' applied both literally and in the sense of the absolute right of advantaged (male) landowners to develop their landholdings as they chose, with little regard for impacts on disenfranchised communities or other classes held by the peasantry bonds of manorialism. In practice, medieval feudalism was a complex system of rights and obligations, within which titled people had privileged rights of access to resources. However, this asymmetry of rights was largely during an era of relatively low settlement densities, and a relative abundance of natural resources per capita. The global human population had not reached half a billion and would not do so until the onset of the Industrial Revolution, which also saw the burgeoning of cities and conurbations.

This model of ownership and access to land and other natural resources was exported widely across the globe wherever European empire-builders took over control, or where their legacies persisted following independence. A particularly extreme, and rather recent, example of the annexation of power and influence by a ruling minority is the apartheid era of South Africa up to 1994. Under the apartheid regime, the ruling white, law-making elite progressively annexed the most productive land and water resources. For example, water constituted the sole property of the owner of the land on which it rose under South Africa's Irrigation Act of 1912, which stated that 'He can do whatsoever he pleases with it and neither the owners of lower-lying land nor even the public can claim to be entitled to make any use at all of that water'. Water represents power and wealth in an arid land, so the net effect of this annexation of water and fertile land was to promote the interests of a dominant class which enjoyed privileged access to resources and economic power, to the detriment of the disenfranchised (predominantly black) majority of the population. As we have seen, the state-protected areas for wildlife, which collectively cover 6 per cent of South Africa's landmass, have a controversial history. Yet private game reserves cover 13 per cent of the land, the implications of which cannot realistically be expected to be any more equitable.[25] Once cemented, the rights of the powerless, landless majority become eroded and eventually progressively forgotten. Eventually, these people ceased to have any voice in governance decisions, and so become excluded by statute, access, education and/ or economic influence.

Elsewhere in the world, nomadic communities have succumbed to the forces of 'sedentarization', which describes the settlement of formerly nomadic pastoralists, as land and crucial resources have passed into private ownership. Forced collectivization under Stalin's rule massively transformed the formerly dominant nomadic herding agriculture of Kazakhstan, undermining an inherently sustainable form of land use adapted to the harsh climate. This collectivization is considered to have contributed to the famine of 1931–34 that may have killed as much as 40 per cent of the Kazakh population.[26] Changing government policies across Egypt, Israel, Libya and the Persian Gulf, in addition to the opportunities of economic development in booming cities responsive to industrialized world culture, have, since the 1950s, effectively led large numbers of traditionally nomadic Bedouin people to become settled citizens of various states rather than stateless nomadic herders, with an estimated decline in nomadic Bedouin from 10 per cent to 1 per cent of the total Arab population.[27] Food crises, droughts and political mandates have contributed to the progressive settlement of many other formerly nomadic communities throughout the past century, drawing them from a direct relationship with the land and into a deepening dependence upon the dominant global economic model.

The fate of both former peasants and nomads illustrates a largely global trend towards the centralization of power and access to key resources in the hands of politically and economically powerful land- and other resource-owning classes. The same trends can be seen in the colonial era of European history, as invading powers – the English in East Africa, the Dutch in the East Indies, the Spanish in Central America and so on – took by force, fenced or otherwise annexed land in their own national, and generally private, interest. The wresting of ownership from local people who had subsisted upon this land prior to European settlement tended to subjugate them to colonial political control, often also to the religious ideologies of the new rulers, and always to the dominant and acquisitive capitalist model.

Capture by the new economic elite

Effectively, and almost right across the globe, there has been a progressive privatization of the land, and an associated transition of the benefits derived from land and landscapes from public goods into private property. Through this process of privatization, the emphasis shifted markedly from communities engaged primarily in activities appropriate to their own subsistence and well-being towards serving the market economy that emerged during the Industrial Revolution. The pattern is repeated in the former colonies of Australia, East Africa, South Africa, the USA, Canada and many other parts of the world, where historic rights and prior claims of nomadic, endemic and other people were largely suppressed by the new political mantra of private owner-ship and the new economic imperative of maximizing short-term profit, generally for private gain, from holdings of land and other resources.

This was not a glorious phase of human history when viewed in ret-rospect. For example, when the Kruger National Park was established in South Africa in 1898, there were mass clearances of tribal people who had lived on the land for uncounted generations, many descend-ants of whom still live in abject poverty in settlement camps adjacent to the park's perimeter. This same pattern is repeated in the USA, where many Native Americans were expelled from land subsequently to be designated as national parks.[28] For example, the establishment of Yellowstone National Park depended on the eviction and killing of hundreds of Native Americans, and also on reneging on previous treaty promises.[29] In addition to this, there are ongoing attempts to move people from protected areas in Thailand, Botswana, Ethiopia, Tanzania, South Africa and India, entailing not only the movement of people but also the destruction of economies and historic associations with the land.[30] It remains commonplace for the costs of conservation to be concentrated on the rural poor of less developed countries, while the benefits are enjoyed by global elites.[31]

The annexation of other resources has an equally inglorious history. South Africa's Irrigation Act of 1912, just one of many regulations transferring control of water, land and other crucial resources from communal stewardship into the hands of a ruling elite under the apartheid regime, has already been quoted. In the USA, Congress passed the now infamous Mining Law in 1872, under which companies and individuals were given the right to buy the mining rights for public land thought to contain minerals for $5 per acre or less regardless of prior occupation or the rights of Native Americans. So too the allocation of mining rights in Australia, entirely overlooking the heritage, spiritual values and prior claims of Aboriginal communities.

The construction of large dams across the tropical and temperate world has also imposed a significant 'take' of land much of which was formerly in common stewardship, often with the disempowerment of many landless people. Somewhere between 40 to 80 million people had been physically displaced by dams worldwide up to the year 2000, the overwhelming majority of whom were already marginalized and some of whom were displaced by violent means, including the shooting and killing of protesters by officials.[32] Official figures have often been found to significantly underestimate the true numbers of people displaced by large dam schemes. Additional people have been displaced by canals, turbines, 'construction villages', compensatory wildlife sanctuaries to mitigate dam impacts, or new settlements created for people displaced by dams, and others displaced by new industries and intensive agriculture exploiting the availability of water and, often, cheap hydropower. Where they have been undertaken, resettlement programmes have generally focused on the process of physical translocation and have almost entirely overlooked 'livelihood displacement', including opportunities for the economic and social development of displaced people together with the social networks that define them and provide them with the necessary resilience to cope with change. Beyond these directly and near-directly affected people, fundamental changes in ecosystems, including for example the regeneration of soil fertility on flood plains downstream of dams, as well as shifts in fish stocks and water-borne diseases caused by disruptions to the hydrology and connectivity of rivers, compromise the ability of many more people to continue to meet their needs or sustain traditional livelihoods.

This brief summary of the historical transition of common to private land ownership shows how this process has served to put power in the hands of the already or newly powerful. This has gifted or favoured a narrow and privileged cadre of society with access to resources that have been denied or withdrawn from the general population. The privatization of land, landscapes and other essential ecosystem resources that occurred through these phases of human history is not without repercussions today, still substantially influencing the assumptions and rights inherited by the modern developed world. For all the advances in public health and personal wealth that this has secured in the developed world, privatization and commercial exploitation of land and other critical ecosystem resources has not necessarily best served the causes of equity and sustainability. Rather than overcoming Hardin's 'tragedy of the commons', excessive centralization of control of resources has instead often served neither to protect nor to allow public benefit from the many functions performed by land and landscapes.

2 | Reclaiming the common good

The progressive annexation of formerly common land into the hands of private owners and interests is, however, only one of the trends in the ownership and control of land and landscapes. As we have seen, common stewardship was widespread before the era of centralized ownership. However, there is also compelling evidence of a steady 're-commoning' of these essential resources in more recent times which has run counter to, and sometimes overturned, governance systems as extreme as South Africa's apartheid regime or the European feudal system and its de facto continuance into the early twentieth century. This chapter explores evidence of significant shifts throughout the past century or so towards renewed recognition of the many public benefits derived from the landscapes within which we live and upon which we depend.

Recovering common benefits

The twentieth century was another period of major, if generally in-cremental, transformation during which the often uncontested land use and development rights of landowners became gradually curbed.[1] Some of this is implicit in the Universal Declaration of Human Rights adopted by the General Assembly of the United Nations in December 1948.[2] The introduction of development planning processes under the Housing, Town Planning Act 1909 and additional guidance on planning control were significant instruments of change in the UK, leading to successive Town and Country Planning Acts from 1947 onwards that sought to restrict changes to land use according to planning policies developed at local and national levels. More recently, under the auspices of the protection of personal freedoms, the Human Rights Act 1998 brought into UK law Article 1 of the First Protocol of the European Convention on Human Rights, securing the individual's rights to enjoy their posses-sions without undue interference from the state. The combination of the two trends is almost contradictory: the Human Rights Act confers an implicit right to develop, which resolves into a freedom for landown-ers to carry out residential, industrial and agricultural development activities, while planning legislation introduces a proviso that these activities should not significantly prejudice the rights and enjoyment of life of others, and it also sets up a regime of democratic accountability for development.

Increasing recognition and protection of the many public benefits stemming from land and its use, regardless of who owns it, relate both to increasing understanding of these broader benefits and how landscapes 'produce' them and also the progressive collapse of the class system throughout the century, which has progressively highlighted the rights of all in society. These examples of UK planning and human rights laws illustrate early stages in a significant and widespread transition during the twentieth century relating to control of land, water and other basic natural resources. They mark the beginning of a transfer from the un-contested hegemony of landowning classes towards recognition of wider societal benefits and impacts flowing from landscapes that may or may not be in private ownership. They also respond to a growing need for a more participative and equitable process for respecting the rights and responsibilities of all constituencies, and not just landowning interests. This pair of examples is neither isolated nor final; rather they serve as indicators of a progressive series of shifts in policy pertaining to the use of land and landscapes, and relating specifically to the balance of private versus public benefit.

There is an extensive academic literature on relative cultural perspectives about the value of land, landscapes, ecosystems and natural resources (reviewed, for example, by Posey,[3] Strang[4] and Ostrom[5]), and of water rights affected by and supporting productive land uses and human activities (for example, van Koppen et al.[6]). We will return to these values later in this book. The purpose of this chapter is not to review that large body of work. Rather, it is to deduce from broader cultural trends evidence of steps towards the attribution of greater weight in public policy towards the rights of all citizens relative to the former centralization of power and rights in private interests. As we shall review in this chapter, many such steps are enacted through legislation and/ or supported by systems of agri-environment subsidies as well as the development of the common law.

Although the primary focus of this chapter is the UK, legal and fiscal examples are drawn from Europe, the USA, Canada and other nations, including, in particular, South Africa. These collectively illustrate a more general trend, albeit that the example nations are all influenced significantly by European governance and economic systems. The South African examples are remarkable in many regards, but particularly with respect to their pace, with pertinent lessons arising from the emergence of the country into democracy from the divisive and domineering apartheid era. This transition brought with it an agenda of reconciliation and redistribution, enshrining the principle of equity over that of hegemony and explicitly acknowledging rights to land, water and other natural resources as central to the societal transformation. Against this political backdrop, substantial reforms have taken place in rights and governance of South African land, water and other natural resources. A significant element of this transition was the patient formulation over a period of three years of what is now the National Water Act 1998[7] with its ambitious vision for Integrated Water Resources Management (IWRM), including a strong focus on the redistribution of water resources towards the poor and the empowerment of historically disadvantaged communities. Rectifying historic abuses and learning about sustainable and equitable uses of both land and water resources are integral to realization of the high ideals of both this Act and South Africa's constitution. As such, South Africa provides a progressive case study of rapid societal transition with respect to access to resources of all types, from which significant lessons may be drawn.

Subsidies for the common good

British and European landscapes have been significantly influenced by subsidy payments for socially preferable patterns of land use over the past century; payments generally tied to management agreements for land designated for specific nature conservation, geoconservation and landscape values. Examples in the UK include Sites of Special Scientific Interest (SSSIs) designated for ecological and/or archaeological value, instigated initially by the National Parks and Access to the Countryside Act 1949. Globally, such designations as World Heritage Sites attract similar protection and, often, favourable subsidies.

Early implementation of incentive and subsidy payments was generally based on the principle of 'profits forgone'. Indeed, this approach is still reinforced by World Trade Organization rules, reflecting the loss of commercial opportunity from withholding environmentally damaging uses of land by private owners. However, with hindsight, this form of 'profits forgone' payment is inherently iniquitous, reinforcing an assumed right of landowners to undertake practices destructive to ecosystems and the many societal benefits that flow from them. The implicit assumption within the 'profits forgone' model is that the public purse should compensate private landowners for forgoing personal profit from practices that undermine the common good. Thus, the feudal system, based on the privilege of landowning classes, remains tacitly intact within this approach to public policy. Furthermore, such initiatives are often poorly targeted and are also reliant upon elective uptake by owners or managers of land.[8] Nevertheless, these early legislative steps represent a progression towards recognizing and safeguarding broader societal benefits, generally pertaining to the cultural benefits (nature conservation, valued landscapes, outflows of fresh water, self-sustaining fisheries, etc.) deriving from land otherwise under the control of private owners.

This marks a tangible step in the transition from the assertion of the rights solely of private beneficiaries. It inherently recognizes broader public benefits that may arise from land use, and is the theoretical basis for agri-environment payment schemes across the European Community. While the EU's Common Agricultural Policy (CAP) has been widely and strongly criticized for its negative impacts upon the environment, at least prior to the 2003 reform,[9] it does provide subsidy streams favouring environmentally advantageous uses of generally privately held land from which nature conservation and other publicly valued benefits may derive.

Particularly since the 1990s, there has been a further transition in the UK towards 'positive management' agreements for agri-environment payments, under which landowners are rewarded not for 'profits forgone' but instead for positive management of land to achieve clearly identified ecological and amenity outcomes. This transition in emphasis from solely private gain towards increasing recognition and safeguarding of societal values arising from land and landscapes is clearly seen in the changing emphasis of the range of land use subsidy schemes that have been implemented in the UK since the 1970s, leading to their harmonization since 2005 in the Environmental Stewardship scheme. SSSI payments in England have become increasingly tied to positive land management,[10] supported by many amendments of the National Parks and Access to the Countryside Act 1949 and its replacement by the Wildlife and Countryside Act 1981 and the Countryside and Rights of Way (CRoW) Act 2000 in England and Wales. Similar legislative changes have taken place north of the border in Scotland and elsewhere across Europe, significantly including the requirements of a suite of environmental legislation emanating from the EU, including the Birds Directive, the Habitats Directive and the Nitrates Directive. All place further protection on the environmental and public benefits of habitat conservation or restriction on groundwater pollution from the use of land generally held in private ownership.

The 2003 reform of the CAP by the EU marked a further progression in land use subsidies, significantly including a shift in the balance of subsidy payments from 'Pillar 1' (market price support) towards 'Pillar

2' (rural development and environmental measures). Ten former major CAP payment schemes were collapsed into one new single payment, with subsidies 'decoupled' from production and instead intended to better acknowledge and reward environmentally friendly and socially beneficial farming practices. In theory at least, farmers would also have greater freedom to farm to the demands of the market and in response to national policies. But also, by explicitly prioritizing rural development and environmental measures over rewards or production for largely private benefit, the 2003 reform marked a transition from benefits for private landowner towards promoting uses of the landscape that deliver public benefits.

For all the welcome change in policy direction under both 'positive management' SSSI and CAP payments (the latter today routed through the Environmental Stewardship scheme in the UK), the positive public benefits for which the subsidies are constructed are often poorly defined. Indeed, there is significant critique of the failure of the subsidy system to deliver clear public benefits, with the UK government itself arguing that large agricultural subsidies are not consistent with sustainable development.[11] This failure revolves substantially around the lack of a common, transparent method to recognize and account for the various forms of societal benefits – both private and public – derived from land and landscapes and enjoyed by all of society.

The USA also operates a substantial agricultural subsidy system as part of a 'farm income stabilization' programme, empowered through a series of Farm Bills that date back as far as the Great Depression in the decade preceding the Second World War, with some legislation (such as the Grain Futures Act) dating back as far as 1922. US subsidies have, of course, shifted over the years as populations have migrated from small farms into cities and agribusiness has grown in individual business unit size to produce often different food and commodities (such as corn and other crops grown for biofuel production). There are wider questions about how well US farm subsidies serve a common as opposed to private interest good, with subsidy programmes providing farmers with extra money for crops and a guaranteed price floor. This secures the important economic role of US farming, but may do little to modify the impacts of agricultural activities on the benefits arising from the land enjoyed by wider sectors of the public. However, given the sheer physical scale of the US landscape compared to a high but, by European standards, low-density population, the tendency is for US environmental subsidies to be targeted at taking land out of production ('land-sparing') rather than to pay for modification of land use. A dramatic departure from this generality, and perhaps a pertinent lesson for the future shape of US farm subsidies, is seen in the case of the Catskills–Delaware water supply system serving New York City, to which we will return in Chapter 8.

Regulation for the common good

In addition to revisions to subsidy systems and their legislative bases illustrated by the UK, EU and US examples above, a similar transition towards safeguarding the health of ecosystems and their many associated public benefits has been witnessed across the globe. The Great Depression in 1930s America is just one example of radical reassessment of the relationship between industrial and economic progress and the people and environment that support it.

The Great Depression was a major shock to the 'American way of life', and perhaps the first such significant shock apparent to the mass population.[12] This may expose inherent fatal flaws in the still-prevalent 'American dream', founded as it is on unconstrained and unsustainable exploitation of land and landscapes, people and other resources. However, some commentators went farther to claim that the Great Depression demonstrated a fundamental failure of capitalism, a sentiment echoed in global stock market crashes and ensuing depressions on a more global scale in the late 1980s and late 2000s.[13] But the capitalist model did not entirely collapse following any of these economic shocks, despite their unfortunate and sometimes serious social repercussions. Rather, each in its own way gave pause for reflection on the need to add ethically and environmentally responsible dimensions to the inherently amoral workings of the market, precipitating various strands of pro-social, pro-environmental and, latterly, pro-sustainable initiatives. The Great Depression was a particularly pertinent moment in American history as it was perhaps the first such 'shock' to an economic model that was assumed to be limitless. Franklin D. Roosevelt's tenure as president from 1933 to 1945 spanned this period of intense societal upheaval, significantly including a raft of 'New Deal' legislation to provide relief for the unemployed, recovery of the economy, and reform of economic and banking systems. Each major piece of legislation saw the establishment of new agencies. This represented a conscious commitment not to marginal policy shifts but to a radical reorientation of national policy, with Roosevelt introducing the change with the words, 'Throughout the nation men and women, forgotten in the political philosophy of the Government, look to us here for guidance and for more equitable opportunity to share in the distribution of national wealth ... I pledge myself to a new deal for the American people. This is more than a political campaign. It is a call to arms.'[14] This radical reorientation, also known as a 'social contract', sought to establish a safety net for the people but also to retrench from a model of exploitation of both human and natural resources that was blind to or unconcerned about its own long-term consequences. This therefore marked emphatically the start of a longer-term cultural response to the social and environmental vulnerabilities becoming progressively more apparent in an inherently blinkered model of industrial and economic progress. It was a forerunner in the dawning of far wider environmental awareness, leading eventually to the Air Pollution Control Act of 1955 (the ancestor of the Clear Air Act of 1970), the passing of the Wilderness Act in 1964 to establish a process for permanently protecting some lands from development, the National Environmental Policy Act of 1969 that saw the establishment of the Environmental Protection Agency, and a host of related legislation sequentially safeguarding aspects of a natural environment that had formerly been seen as fair game for industrial entrepreneurs.

America and Canada have also seen a reassertion of the rights of aboriginal peoples formerly disenfranchised from their land by predomi- nantly European settlers labouring under the delusion of a 'Manifest Destiny'. This loosely constructed concept of the superiority of the Euro- pean race (itself of decidedly hybrid provenance) was taken as a mandate to occupy land in a westward expansion and to contribute to an 'Indian Removal', or to bring to these people the benefits of Western-style 'civili- zation' that was little more than the imposition of European capitalism and religion upon a system of communal resource sharing that had existed for millennia. This negative attitude to the rights of aboriginal people persisted through to 1924, when Native Americans who were not already US citizens were granted citizenship by the Congress of the United States, with the 1975 Indian Self-Determination and Education Assistance Act also representing a culmination of fifteen years of policy changes that formed part of the wider civil rights movement. Parallel developments in Canada were enjoyed by legally defined First Nations people whose rights became recognized with various land redistribution initiatives to remedy some of the many historic 'wrongs'.

Various land claim processes across other parts of the world have begun to rectify some historic injustices with respect to the disenfran- chisement of aboriginal people and their clearance from their former homelands, including examples in Australia and various African coun- tries. While the intent of these initiatives is the reallocation of land out of the hands of private owners to the races or tribes from which it was historically stolen, the ways in which land redistribution has occurred across Africa are wildly uneven. They include seizures of white-owned land in Zimbabwe and its redistribution by patronage under the dictatorial government of Robert Mugabe, peacefully negotiated land claims in neighbouring South Africa, and enforced but largely non-violent 'land claims' in Kenya based on restoration of the rights of tribal people to land seized during historic European colonialism.

The opening of the UK landscape to public enjoyment through various Acts of Parliament since the Second World War, extending from the National Parks and Access to the Countryside Act 1949 through to the 'right to roam' clauses of the CRoW Act 2000, also represents regulation to protect the 'common goods' provided by land. The CRoW Act also set up the statutory English nature conservation agency Natural England, established with the purpose of ensuring '... that the natural environment is conserved, enhanced and managed for the benefit of present and future generations, thereby contributing to sustainable development'. This unambiguous link to human benefits further enshrines the protection and promotion of public benefits arising from land and landscapes, regardless of land ownership status.

Notwithstanding continuing difficulties with the implementation of its visionary laws,[15] South Africa's rapid and radical transition in legislation regarding land, water and other natural resources still remains an exemplar of change at national policy level to transfer power from a model of hegemony to one of redistribution that respects the rights of all in society. The dissolution of former land and water rights and their reapportionment under interest-based negotiation, beyond a minimum level secured by the state to support basic human needs and the maintenance of ecosystem functioning, is a principle with generic applicability across the world.

Progressive constraint of the power of the corporate sector through various conventions, protocols, market pressures, environmental and social regulations and corporate governance requirements also reflect a rebalancing of private interests in natural resources relative to public enjoyment. Multinational corporations still have immense influence, some of them more economically powerful than many small nations,[16] yet the process of recognizing their potential societal benefits over and above private profit alone has been progressive, with business sometimes seen as a key player in problems but also their solutions in a world facing daunting sustainability challenges.[17]

Common law for the common good

Common law, or case law, differs from the statute law in that it evolves through successive judgments, rather than through legislative statute. Common law judgments often relate to interpretation of the principle of 'fairness'. As such, the common law has been a significant instrument for addressing the rights of all who benefit from the use and functions of land since its origins in Europe in the Middle Ages, with some strands dating back to Roman Justinian Law, and since further developed in many nations that trace their legal heritage to former European colonies. The common law evolves far more rapidly than its statute counterpart through case law, reflecting shifts in knowledge and public values with respect to basic principles of human rights. Improved understanding of the interdependence between ecosystems and society has been significant in the progressive evolution of common law with respect to environmental rights.

Just as statute law has evolved to redress the balance between private and public benefits stemming from land and its uses, the common law has also been in transition. This includes increasing recognition of the rights of public beneficiaries from the processes performed by landscapes, including culturally valued places, air purification, storage and cleansing of fresh water, biodiversity and recreational opportunities, all of which may be affected by the actions of the owners and managers of land and other resources.[18] Some of the most substantial case law relates to redress for damages suffered by people displaced by the construction of large dams in a number of countries, particularly in the USA and Canada. Evolving civil rights have resulted in many examples of displaced communities seeking retrospective compensation for the consequences of dam-building and dam-filling, the commissioning of which was initiated in less enlightened times before the rights of less privileged sectors of society were recognized. Two particular examples illustrate successful common law cases entailing substantial compensation reflecting significant historic damage to parties not formerly deemed worthy of involvement in decisions pertaining to dam schemes.

The first example is the Grand Coulee Dam on the Columbia river in the United States. Developed between 1933 and 1955 for hydropower generation and irrigation supply, this massive dam was conceived in an era when impacts on the migration of fish, particularly salmon, and the consequences for the livelihoods of upstream Native Americans and Canadian First Nations were, at best, inadequately considered. Production of salmon and other fish had been the centrepiece of the area's indigenous economy and culture. In 1951, the Colville Confederated

Tribes filed a suit against the United States. Twenty-seven years after the claim had been filed, the Indian Claims Commission ruled in 1978 that the tribes were entitled to full compensation for all income losses associated with the dam, for which the US government provided a total of US$66 million as historic compensation, including annual payments of US$15 million to offset ongoing reduced income opportunities. Cash alone, of course, cannot meaningfully compensate for the degradation of cultural heritage and many of the irreplaceable supportive capacities provided by ecosystems.

The second example is a similar case revolving around impacts on the locally adapted livelihoods and lifestyles of indigenous people in the James Bay area of Canada. When commissioned in 1970, the various dams and reservoirs of the James Bay hydropower complex in Canada did not provide any specific compensation for indigenous populations, who nevertheless suffered significant social impacts, including in particular those arising from reduced incomes from hunting and fishing due to hindrances to the free migration of fish and loss of breeding grounds of geese and caribou. The indigenous people initiated legal action on the basis of unsettled land claims, forcing the James Bay and Northern Quebec Agreement, signed in 1975, which provided compensation in the form of a 'remedy fund', including a comprehensive lump sum and an annual payment to address outstanding social issues.

These are just two among many common law cases which have retrospectively redressed damages to stakeholders excluded from decision-making about resource exploitation purportedly for the public good. Societal value systems eventually evolved to a point that enabled injustices to be remedied under the common law, and which in their own ways substantially shaped the subsequent evolution of statute law.

Rights in transition

The progressive transition back from the hegemony of landowning classes towards recognition of the 'common good' benefits of landscapes is far from complete. World Trade Organization rules still enforce the 'profits forgone' basis of land use subsidies. Yet an indication of the ongoing shift in public opinion towards the rights of all in society is the media attention devoted to many instances of actions taken in the interests of private people or enterprises that are in conflict with the 'public good'. Practical examples here include human rights and employment law conflicts, abuse of parliamentary privileges by MPs and, as still commonly seen in local media, disquiet about seemingly unjust development planning decisions, including the abuse of privilege by elected officials.

In this context, commonly held rights to land and other natural resources, whether or not the resources themselves are in public or private ownership, do not automatically imply that they will be subject to Hardin's 'tragedy of the commons'.[19] 'The tragedy of the commons' describes a tendency towards over-exploitation of resources held in common ownership due to individuals having an incentive to maximize the benefits they take from the 'common' through resource exploitation, but with the associated costs borne by others sharing the resource. According to the metaphor, without collective stewardship this will eventually lead to the progressive over-exploitation and destruction of common resources. Significant evidence of this outcome can indeed be seen in the collapse of international marine fisheries, tropical forests, aquifers and other 'common' ecosystems under competitive commercial exploitation.[20] Building upon this tendency is also 'the tyranny of small decisions', wherein the cumulative impacts of incremental developments, seemingly individually of low significance, cumulatively exert substantial and potentially catastrophic pressures on shared natural resources.[21] This creates problems for governments and societies across the world, from a proliferation of small wells that cumulatively deplete aquifers to myriad small instances of forest clearance, abstraction from surface water resources, 'take' of fish or other wildlife from dwindling populations, or localized small emissions of greenhouse gases that contribute to major diffuse contributions to climate change.

However, we would be unwise to assume that 'the tragedy of the commons' is an automatic consequence of the common ownership of natural resources. In fact, stewardship of land in common ownership, both prior to eras of privatized ownership and continuing today, is often found to be backed up by social infrastructure or lifestyles adapted to the landscape's regenerative capacity. With appropriate governance and cooperation, management of common land can be entirely viable and sustainable, with livelihoods adapted both to the carrying capacity of water, soil and other natural resources and backed up by appropriately adapted cooperative social structures. Common principles have been found to underlie many such enduring common resource systems across different cultures and nations.[22]

A prerequisite of sound and enduring stewardship is that the interests of multiple constituencies of beneficiaries are reflected in management practice. There is a role for both common and statute law in the recognition and restoration of wider public rights where these have been historically appropriated into private ownership, consistent with the principle of inclusivity. 'Inclusivity' has various definitions,[23] but can be summarized for this purpose as giving a voice to all interested parties in policy- and decision-making. If we are moving from a situation of dominant private interests towards respect for public interests, the ways that we go about framing policies and legislation, and arriving at ensuing decisions, must equally be subsumed within wider participative processes.[24] Recognition of public rights must necessarily mean that we have to accommodate broader stakeholders in decisions at all scales, rather than staying loyal to our historic dependence upon the views of a few selected 'experts' who often reflect relatively narrow political or special interest perspectives, effectively acting, albeit often unintentionally, as a new class of feudal 'lords'. Notwithstanding its democratic intent, this outcome is frequently seen in executive, as opposed to participatory, governance, such as is seen in British parliamentary democracy. We have to find ways of bringing other perspectives and types of knowledge into the development of public policy, law and benefit valuation.

The need for truly inclusive dialogue is central to the transition of water management in South Africa from its apartheid-era rights basis to a future interest basis for allocation. A key aspect of the South African paradigm shift entails recognition that the allocation of entitlement to use of water is essentially a social process of bargaining. Rights-based and interest-based processes, well documented in the industrial stakeholder relations literature and increasingly in the natural resources management literature,[25] are markedly different in character and have implications for the human, financial, legal and technological resources committed to the process. Significantly, this necessarily entails creation of consensual frameworks, and possibly also revisions of institutional arrangements to accommodate them.[26] However, we have historically lacked a common and transparent basis for recognizing, evaluating and sharing the multiple benefits derived by society from land and landscapes. This conceptual gap has thwarted aspirations for a consistent approach to the sustainable and equitable use of land, akin to our observations about the progressive evolution of land use subsidy systems.

Governance for the common good

The transition in the societal appreciation and valuation of land in the UK and Europe, South Africa, North America and elsewhere has not occurred in isolation, but within a broader international context of growing environmental awareness and a rising realization of the dependence of human well-being upon ecosystems.[27] Key elements of the evolution of environmental awareness and the development and engagement with environmentalism and human equity within international discourse during the past century have been extensively reviewed elsewhere.[28] Suffice to say that they have seeded modern conceptions of sustainable development, integrating the needs and vitality of supporting ecosystems with economic and social progress. In essence, this modern conception of sustainable development regards ecosystems and their supportive capacities as a property and responsibility of society as a whole, with the need for integrated decision-making that simultaneously respects the integrity of ecosystems, people and the economy, including the rights of all present and future stakeholders.

The concept of sustainable development is not without its critics. Some argue, for example, that it is pro-development, excluding debate about the 'deep green' appreciation of the inherent worth of ecosystems, and that this automatically prioritizes the utility and exploitation of natural resources and the pursuit of economic progress over their intrinsic worth.[29] A further argument is that, since the Earth's resources are finite, the concept of 'sustainable economic growth' is an oxymoron,[30] and that monetization is anyhow necessarily too imprecise.[31] Others make a similar observation about developed-world economic priorities and perspectives potentially overriding the rights and value systems of developing nations and cultural minorities, particularly for those people outside of the 'first world' economy.[32] However, the very purpose of sustainable development is not the prioritization of one world-view over others, be that economically, socially or ecologically founded, but rather their integration within a model that seeks the simultaneous protection or advancement of all three strands.[33] To argue that the inherent worth of ecosystems should put them beyond economic use or the support of societal need is to believe that the economy can be set aside, which is neither realistic nor does it acknowledge the interdependence of human and ecological systems. Instead, sustainable development demands radical reform of the market, human relationships and resource use habits consistent with protecting what is held to be of ecological and cultural importance and adapts to the 'carrying capacity' of global ecosystems.[34] This is a radical transformation indeed, making space for multiple value systems and perspectives within a development model that can not only continue indefinitely but is equitable to all in society and preserves its ecological bedrock.

The challenge of sustainable development therefore implicitly necessitates a participative approach to decision-making that respects the rights of all of humanity, acknowledging that all are ultimately supported by common ecological processes, significantly including those performed by land and landscapes. This, in turn, suggests a fundamental shift away from largely uncontested private rights to develop land, and the dominant economic model beneath it which generally overlooks the dependence and impact of commerce upon ecosystems and the rest of society. We should expect the law to react to these changing conceptual and moral perspectives, as well as to various international protocols and binding commitments.

One such example was the UN Aarhus Convention,[35] adopted in June 1998, as a new kind of multinational environmental agreement explicitly linking environmental and human rights. The Aarhus Convention acknowledges that we owe an obligation to future generations, which can only be protected by reflecting the views of many stakeholders in decisions relating to the environment. Binding requirements of the Aarhus Convention are embodied in recent EU Directives such as the 2000 Water Framework Directive, requiring wider public engagement in deliberation from the problem identification and framing stages through to options generation and appraisal, identification of solutions and their refinement, and into implementation and subsequent monitoring of management activities. This is at significant odds with the established 'consultation' practices, often characterized by the polling of reactions to decisions, but in which the final decisions are made by those who are doing the consulting.[36] This tends to favour 'expert'-driven decisions often filtered by sunk cost in design and other considerations, vested interests and lack of consideration of alternatives, often entrenching a 'favoured option' as a foregone conclusion. Conversely, broad stakeholder participation is generally now seen as leading to better and more sustainable decision-making.[37] Although the Aarhus Convention has been statutory for some time, and is intended to apply to the implementation of previously established legislation in signatory countries (including the UK via the EU, as well as Canada and the USA), there remains low awareness of the Convention and a serious shortfall in its implementation. Its requirements for comprehensive public engagement and deliberation, informing eventual decisions, remain far from widely in evidence today.

The challenge of governance is to recognize that the needs of all, now and into the future, depend upon a common pool of natural resources which, though many are inherently renewable, have finite limits beyond which their integrity and supportive capacities are compromised. It is therefore inevitable that disproportionate demands by one sector of society will affect the capacity of ecosystems to support the well-being, economic opportunity and security of all other sectors, including generations as yet unborn. A common language around which to frame common understandings, and to negotiate and govern on the basis of sound communal resource stewardship, is vital if we are to aspire to sustainable and equitable outcomes. Clearly, this runs counter to historic and inherited assumptions rooted in the annexation of land and influence by powerful oligarchs, as manifested, for example, in European nations that formerly established empires across the rest of the globe. However, it may not be incompatible with a private land ownership model given the right checks, balances and governance systems to respect the benefits and rights of many sectors of the public supported by the functions of that land.

Reclaiming public goods

Society is today making rapid progress with novel tools to connect people with natural resources and to plan for their more sustainable sharing at a 'grassroots' level, increasingly recognizing the fundamental importance of ecosystem health to public and economic well-being. For example, the UK's Natural Economy Northwest (NENW) project[38] was an innovative three-year programme, initially set up to run to December 2009 and to involve many regional partners, promoted by the Government Office for the North West, which focuses on the 'green spaces' and 'blue spaces' between urban centres in the north-west of England. The aim of NENW was to maximize the benefits of investment in the region's natural environment, both underpinning and delivering sustainable economic growth. Other government regions and Regional Development Agencies across the UK also highlight the central importance of the natural environment to the well-being and character of their regions, as well as their role as a fundamental base for future economic progress. A variety of related tools and initiatives to address the relationship between urban development and supportive ecosystems is now emerging both in the UK and globally, including, for example, 'green infrastructure' (www.greeninfrastructure.co.uk), sustainable drainage systems or SuDS (as reviewed by Woods et al.[39]), 'green lungs' and 'wildlife corridors', the National Community Forests Partnership (www.communityforest.org.uk), Eco Cities (www.ecocitiesproject.org.uk), the Indian GIST programme (Green Indian States Trust: www.gistindia.org) which recognizes that 'Natural Infrastructure must be treated the same as built infrastructure: invest in it, account for depreciation ... a must for sustainable development', and many more besides, including some with a specific focus such as climate change initiatives under the Nottingham Declaration (www.energysavingtrust.org.uk/nottingham). All of these apparently disparate initiatives are founded on placing ecosystems and various of their capacities and services centrally in urban planning and development. At present, their innovation and implementation are fragmented, addressing different, entirely valid and important aspects of the wider relationship between urban development and ecosystems. Yet all can be contextualized within the breadth of connections between ecosystems, social needs and economic opportunity.[40]

Notwithstanding the pervasion and strength of this observed shift from predominantly private advantage to wider consideration of enduring value to the public, change seems to have happened on a largely ad hoc basis, lacking a cohesive framework for understanding and planning. Yet we may be witnessing a deeper systemic change in beliefs and structures.[41] However, what is conspicuously lacking is a consistent basis for the sharing of the most vital of all resources upon which society depends, and which we need to learn to share wisely and equitably: the ecosystems that support all facets of human needs and security. These include crucial assets such as water, air, land and landscapes, which provide people with the most fundamental of resources to live biologically, culturally and economically fulfilled lives. Collective recognition, valuation and stewardship of these resources is integral to society's progress with equitable and sustainable development. The process of 'reclaiming' these as public goods, reversing the acquisitive privatization tendencies exhibited over prior centuries, has been a striking trend of the twentieth century and beyond.

To support and accelerate this transition, we need to develop conceptual frameworks and pragmatic tools based upon the properties of the ecosystems that we share. This will enable us better to work together towards securing both these critical natural resources and the best interests of all who share and depend upon them.

3 | The ends of the Earth

Privileged access to resources would be immaterial if the Earth's ecosystems were infinite. However, natural laws set finite limits to their capacity for regeneration, and many are already exploited beyond those limits. This compromises not only our ability to share them equitably but also the integrity and resilience of the ecosystems themselves. Contemporary human resource use habits are already at the ends, or beyond, of the Earth.

Global ecosystems function as a fully interdependent whole, including the sum total of human metabolic, land use, industrial and other activities. Through our unique capacities to comprehend and innovate, we have manipulated the Earth's resources on a scale unprecedented in evolutionary history. The cumulative consequences of this often excessive use of the ecosystems that sustain us are all too familiar today. Runaway trends in human population and rising material expectation among less privileged nations that are 'catching up' with the profligate habits of the already developed world exacerbate current pressures. The disconnection of modern industrialized society from the landscapes that support its most basic needs, blinded by an economic model that undervalues resources of fundamental importance, compounds its vulnerability to over-consumption. This 'perfect storm' of pressures threatens to snuff out further human progress unless we make drastic and rapid changes to contemporary lifestyles and habits.

The term 'carrying capacity' appears to have been coined in the first half of the nineteenth century to define the supportive limits of ecosystems beyond which their breakdown or degradation becomes inevitable.[1] Yet we seem, in hindsight, to have been sparing in the use of our advanced faculty for foreseeing the consequences of our actions, overlooking the largely predictable consequences of overriding the limits of these natural systems. Today, we continue, quite literally, to go beyond 'the ends of the Earth' by exploiting ecosystems well beyond their carrying capacity. There is a pressing need to reorient our planning to address long-term viability rather than merely short-term profit, with ecosystem resources brought centrally into decision-making.

Overriding nature's limits

It seems that humanity has always had the need for an apocalyptic archetype. Our history and literature are littered with fearful characterizations, from the horsemen of the Apocalypse to communist conspiracies, the 'shadow of the bomb' that lurked malevolently over the childhoods of those growing up in the 1960s and 1970s, the groundless demonization of races or religions, demonic images such as Satan, and so on. But today's threats are no chimeras. Instead, we have hard, scientifically based and tested substantiation that the trajectory of our industrial lifestyles has eroded the roots of the 'tree of life' that supports us. Abundant evidence for this is manifest in declining quantities and qualities of fish stocks and old grown forest, loss of water-trapping habitat in uplands and flood-attenuating plains and wetlands lower in catchments, eroding soil and oxidization of its carbon content, persistent naturally occurring and man-made pollutants, degradation of treasured landscapes, climatic instability and food security, and depleting biodiversity, stocks of natural resources and a range of other measurable environmental variables, all of which are already drastically eroding human well-being, enjoyment and potential.

The most globally consistent such analysis is that of the UN Millennium Ecosystem Assessment[2] (the MA). The MA was certainly authoritative, developed by over 1,300 scientists across 95 countries and supported by the consensus of many thousands more in other nations besides. It provided unambiguous evidence that humanity is impacting all major ecosystem types across the planet to an alarming rate. The prognosis for human well-being is stark and disquieting. The findings of the MA do not stand alone but confirm and build upon long-standing studies such as the Worldwide Fund for Nature's (WWF's) excellent series of *Living Planet Index* reports[3] published since 1990, which catalogue the decline of global ecosystems and the spiralling demands of modern society upon them. They also add to the findings of the long-running (also since 1990) series of UNDP *Human Development Reports*,[4] which clearly describe the impacts of inequitable sharing of wealth and power, and the implications of systematic environmental degradation for peoples both poor and rich.

These studies endorse such authoritative expressions of concern as the World Scientists' Warning to Humanity, published on 18 November 1992, just months after the 'Earth Summit' in Rio de Janiero and signed by more than sixteen hundred senior scientists from seventy-one countries, including over half of all Nobel Prize winners. This warning did not mince its words, noting that 'Human beings and the natural world are on a collision course ... A great change in our stewardship of the Earth and the life on it is required if vast human misery is to be avoided and our global home on this planet is not to be irretrievably mutilated ... No more than one or a few decades remain before the chance to avert the threats we now confront will be lost, and the prospects for humanity immeasurably diminished.' Not words to inspire complacency, nor be dismissed as the conclusions of uninformed reactionaries. Further global consensus about the need for a significant change in the direction of human development is provided by the unambiguous conclusions of the Intergovernmental Panel on Climate Change (IPCC)[5] that we are already witnessing the impacts of climate change, which will continue to intensify owing to current let alone future gaseous discharges from human activities, and that we have perhaps ten years to make serious reductions in our carbon emissions if rapid and potentially irreversible climate change effects are to be averted.

This manifests directly in the economy through the costs of timber, food and clean water, damage from increasingly frequent and intense storms, the declining reliability of production from extractive and agricultural industries, and difficulties securing loans and planning permission. All of these pressures and more besides are battering our pension funds and the wider economy.

Rediscovering the Earth's finite supportive capacities

There is today an unprecedented need to marshal our unique capacities for foresight, innovation and collective action to assure ourselves of a future. Sustainable development has become the defining challenge of our age, notwithstanding some of the criticisms of it which have been discussed in the preceding chapter. We have much to learn about the future, but also a wealth of lessons from past civilizations that have failed to heed warnings stemming from depletion of fundamental ecological resources.

Jared Diamond's seminal book *Collapse*[6] considers many examples of societies that have overlooked or ignored the vitality of core underpinning resources from soil fertility to forest resources, wildlife to water, and spanning the tropics to polar regions. Diamond distils a set of generic driving principles for such societal collapses, some of which are total and catastrophic, others more gradual, and some even averted by luck or by sound governance choices. We can add to Diamond's examples the 'dust bowl' of the American Midwest, the precipitous crash of the Grand Banks cod fishery of the northern Atlantic, the massive loss of topsoil and contribution to the devastating flood of 1998 resulting from forest clearances in the upper Yangtze river basin in China, and many more instances besides of ecosystem degradation undermining human livelihoods and potential.

Recognition of the pressing need for the safeguarding and equitable sharing of ecosystem resources underpins much of the long and complex story of the rise of 'green' awareness, leading to modern conceptions of sustainable development. Simple though the integrating principle of sustainable development may be, practical action in a world of deeply entrenched habits, assumptions and vested interests is a more complex process. Yet challenges remain to make these concepts tractable and operational in the day-to-day decisions that cumulatively shape a different future, and which may overturn vested interests in favour of a collective 'common good'. This requires clarity about the challenge and its interpretation, though there is evidence of attempts at obfuscation. For example, in the two years subsequent to publication of the 'Brundtland Report', around 140 alternative and variously modified definitions of 'sustainable development' emerged with perhaps 300 such definitions of 'sustainability' and 'sustainable development' proliferating by 2007.[7] Seemingly, these modified interpretations of sustainable development were the products of vested interests, serving a variety of agenda, most of which do not necessarily have at their heart the well-being of the planet's supportive ecosystems nor that of people in the developing world or future generations.

Sharing the global commons

Lessons learned about the importance of cooperation and sound governance for the sharing of a common pool of resources magnify from the village green or pond, allotment or other patch of land managed communally for the production of food and fibre. We are instead dealing today with international oil reserves and intercontinental migratory waterbird flyways, global forest and peatland resources scattered across many countries that cumulatively provide a crucial bulkhead against runaway climate change, planetary-scale climatic stability and the integrity of trans-frontier rivers that respect no political boundaries. The keys to our expectation of living rewarding lives in future lie not in yesterday's model of 'winner takes all' annexation of common biospheric resources. Rather, they reside in how we recognize, value and learn to share land, landscapes, water and the other vital resources that are inescapably our common foundation and legacy. Rather than concern ourselves merely with shared grazing, our preoccupation today needs to be with sharing of the 'common' of the vitality and productivity of the

Earth's entire biosphere. This is both the sole common habitat for all of humanity and the wellspring of our collective future. Our busy, populous human lives have already reached out to the ends of the Earth and are threatening tangibly to consume the vital resources that underwrite both its and our prospects for the future.

We have an urgent need for common and transparent understanding of the finer capacities of ecosystems and our cumulative pressures upon them. We also need practical methods for recognizing and accounting for the many benefits conferred by land and landscapes upon all of society. We then need to acknowledge those limits and to reflect them in our economy and resource use habits. We will turn our attention to redressing this deficit in the following chapter.

4 | Shifting conceptual landscapes

Throughout our journey of increasing technological and cultural sophistication, the ways in which we have understood, explored and managed the landscapes and broader environment that provide for our needs have evolved through a number of conceptual shifts. All of these have significant implications for the vitality of those ecosystems and their capacities to support our long-term interests. It is instructive to explore some of the different world-views, and transitions between them, that have informed successive recent phases of human advancement. This is valuable not merely to understand better the inherited assumptions with which we live, but also the kind of understandings and actions that will be necessary to live in a more secure and sustainable future.

In particular, the ways that these different world-views influence the natural environment and our relationship with it are a vital determinant of our future prospects. Conservation of the diversity and scale of ecosystems remains an absolute priority if we are not to continue to undermine their capacities to support human opportunity in ways that are well, poorly or not understood, both now and into the future.[1]

Paradigms of human understanding and action

Some of the things that we learn help us understand the world better. To take an ecological example, our understanding of ecosystem structure and function is enhanced by the sciences of taxonomy and evolution, food webs that efficiently cycle energy and matter, and the interactions between living and non-living components within complex ecosystems.

Other things that we learn help us understand the world not so much better as differently. For example, knowledge about the dilution capacity of rivers helped society disperse its wastes until the point at which population boomed in growing cities. New knowledge about the impacts of organic wastes on river quality and of the transmission of diseases through polluted water spawned innovative waste-water diversion and treatment technologies, breaking down the formerly perceived truism that 'the solution to pollution is dilution' when inputs of substances overload natural breakdown and assimilation processes. Further knowledge revealed the nascent threats of bioaccumulation, endocrine disruption at far lower concentrations than formerly accepted 'safety' levels, and of climate change, including the contribution of energy-intensive waste-water treatment processes to rising global temperatures and the need to innovate a new generation of treatment technologies that are benign to humanity and ecosystems. These sequentially different types of knowledge each challenged and ultimately overturned the assumptions behind pre-existing knowledge frameworks, and also the perceived wisdom and efficacy of associated management solutions based upon them.

The evolution of human understanding is, then, not merely additive but can be revolutionary, progressing through different world-views and perceptions of mechanisms, problems and opportunities. Thomas Kuhn's seminal 1962 book *The Structure of Scientific Revolutions*[2] coined the term 'paradigm' to describe world-views that provide agreed conceptual frameworks for assumptions against which to marshal specialist knowledge and inform decisions. Kuhn observed that humanity has passed through a number of paradigms. 'Normal science' describes accumulation of knowledge consistent with an established paradigm. By contrast, 'paradigm shifts', also referred to as 'extraordinary science' or 'revolutionary science', describe science that challenges or changes basic assumptions, ultimately leading to new paradigms. Different world-views lead us to different awareness, analyses, technologies and outcomes. Today, it is generally accepted that three of the most significant paradigms through which science has developed include reductive science, holism and systems thinking.[3]

While Kuhn's observations related exclusively to science, the concept of paradigms has since been widely applied to public policy, management and other aspects of human activity. The following sections of this chapter consider environmental management through the 'lens' of the three paradigms of reductive science, holism and systems thinking, and the shifts between them. They are selected for further exploration as each has proved particularly significant in understanding and managing the modern world from the European Industrial Revolution onwards, and projecting out into the future.

There is a bias in the examples of each towards freshwater ecosystems and the catchment landscapes that generate them. The reason for this is that fresh water is frequently a limiting resource for human development around the world and, as a consequence, freshwater ecosystems and socio-economic relationships with the catchment landscapes in which they lie are well studied. Furthermore, all land is a constituent element of drainage basins, intercepting, recycling and transforming water flows and quality, and so their associated freshwater ecosystems integrate multiple natural influences and human pressures across broad landscapes. As such, they are particularly vulnerable to human-induced stresses, with the biodiversity of freshwater wetlands highly susceptible to these integrated pressures.[4] For example, the prognosis of current trends for freshwater fish gives cause for serious concern as 200 (38 per cent) of the 522 European freshwater fish species are threatened with extinction with a further twelve already extinct, representing a much higher level of threat than that facing either Europe's birds or mammals.[5] As a consequence, freshwater ecosystems worldwide are in decline, along with their capacity to support human well-being.[6]

The world view of reductionism

The paradigm of reductive science, or reductionism, entails reducing complex systems to their component parts as a means for understanding and exploitation. It formed the basis for Sir Isaac's Newton revelatory discoveries about mechanics and their wider application to understanding the workings of the universe. Analysis of constituent parts of natural systems proved a major spur to human understanding, for example in medicine by looking at discrete organs, tissues, cells or biochemicals, and in engineering by observing the mechanics of combustion and applying that to combustion engines. Reductive analysis of the world substantially stimulated the innovations of the Industrial Revolution, approximately from the late eighteenth to the late nineteenth centuries, so has had a profound influence in shaping the modern world, its assumptions, technologies and institutions. Reductionism proved, and continues to prove, one of the most powerful explanatory tools ever developed.[7]

The application of reductive science created previously unimagined capacities to harness the energy of fossil fuels and river flows, to drain landscapes and erect walls against destructive flooding, to build edifices, and to turn raw resources into useful and profitable products. This drove industrializing societies increasingly to exploit resources to advance technology, science, art and philosophy, international trade, cultural exchange, empire-building, public health, material expectations, at least among advantaged sectors of society, and the generation of personal wealth. The innovations of the industrial age were without precedent.

However, the narrow, problem-solving focus of reductionism does not recognize broader interconnections, including the potential for longer-term consequences and competing interests. Land and landscapes, water resources and their associated ecosystems are instead largely viewed as exploitable assets, implicitly considered limitless in their capacities for renewal and waste assimilation. Potential ramifications such as unintended pollution, resource depletion, social equity and other aspects of the sustainability challenges with which we are increasingly familiar today were largely overlooked. Furthermore, as exemplified by the extreme example of South Africa's Irrigation Act of 1912 (see Chapter 1), ecosystems became construed as the 'property' of landowners,[8] whether for trade or the control of other people. This restricted perspective of the natural world as an exploitable resource has significant implications for ecosystem integrity and the sharing of natural resources.

A particular anomaly from the Georgian and Victorian eras is seen in many surviving artefacts, particularly preserved biological collections, museums, natural history societies and the Romantic periods of both poetry and painting. These reveal the fascination of our industrial forebears with natural history. While this fashion for preserved collections may have driven some rare, and hence desirable, species into greater scarcity and even extinction,[9] the greater irony is that the industrial enterprises and resource use patterns generating the wealth that funded scientific expeditions, new institutions and collections were to become the chief agents of resource exploitation habits that remain instrumental today in ecosystem decline.

This narrow perspective of land as a resource for building and farming led to widespread drainage and 'reclamation' of land prone to flooding. This precipitated the invention and proliferation of flood banks and other hard-engineered 'flood defences' designed to hold back flooding, which was generally conceived as an unpredictable 'Act of God'.[10] Flood hazard research then was encapsulated in a perception of 'environment as hazard',[11] focused on land drainage as a predominantly profit-seeking private enterprise financed by its direct beneficiaries.[12] This 'hard engineering' approach was undertaken with no thought for the concentration of flood peaks downstream, degradation of nature conservation and fishery interests, loss of natural processes purifying air and water, and associated landscape and amenity values.[13] Water-dependent ecosystems and people reliant upon them were, instead, largely overlooked.[14]

Decision-making by those with privileged access to land and water also sometimes involved a degree of collective action to advance private interests. In South Africa, Irrigation Boards were established for the collaborative benefits of private-interest water users. In the UK, various non-statutory organizations were formed, some becoming enshrined in statute law. For example, the establishment of Internal Drainage Boards (IDBs) by consortia of landowners collectively managing the drainage of land dates back to 1252, although most current IDBs were reconstituted by national government following the Land Drainage Act 1930. Collectives of water-controlling landowners were also established in lowland England from the medieval era to develop and control management structures in rivers to harness and divert flows to serve cascades of water meadows, watermills and navigation interests.[15] The emphasis of UK flood management during much of the twentieth century was upon land drainage,[16] and this was transferred from private right to public duty by the Land Drainage Act 1930, which established legal meanings for the concepts of flooding and 'flood defence' related largely to lowland agriculture and land drainage for the national economic and military interest.[17] This served to transfer the costs of preventing flooding and draining land from landowners to general taxpayers. 'Reclamation' of productive farmland from coasts and flood plains and their subsequent 'defence' was substantially promoted in the 'Dig for Victory!' initiative (discussed in greater depth in Chapter 6) during and following the Second World War, as self-sufficiency in food production became a political priority. This further reinforced a reductive view of land as largely a medium for growing food.

Recognition of unintended consequences for nature

As industrialized approaches to land and other resource use continued to shape the modern, ever more populated world, the consequences for the natural world began increasingly to be recognized – for example, in the seminal book *Man and Nature*, published by the American George Perkins Marsh in 1864.[18] Although much of the history of nature conservation was framed under the holistic paradigm that was to follow, when interdisciplinary interactions became far better understood, early responses to concerns about declining wildlife included establishment of nature reserves reinforcing the reductive view that areas of importance for wildlife, flora, fauna or features of geological or other special interest were literally as well as conceptually separated from other human activities. As described previously, the antecedents of some strands of conservation policy can be found in the founding and management practices of princely India, with all the privilege and marginalization that that implies. In 1864, Yosemite Valley in the US state of California became established as a preserved landscape, albeit initiated by supporters largely comprising wealthy sportsmen and outdoorsmen, with Yellowstone National Park established in the US state of Wyoming in

1872. Similar nature reserves as well as 'wilderness areas' progressively became established in pockets around the world, either at the instigation of governments and their institutions, by private landowners including charities, research institutions or special interests such as game or fishing reserves, or in the form of ancient sacred areas in which access or hunting was prohibited.

This narrow and elitist approach to conserving nature perceived as threatened or of sporting or other interest was consistent with the dominant property-owning and privatized rights model under reductionism, with decision-making centred on the interests of the resource-controlling and privileged minority.

The reductive world-view shaped many elements of the modern world, including an economic system founded on the utilitarian value and profitable liquidation of natural resources and minimization of costs associated with human resources and waste disposal. This, together with systems of corporate governance and the creed of capitalism which has since become more globally pervasive than any faith or political system, remains in effect today and is the cause of many of the environmental and social equity problems with which we are now grappling.

The reductive model dominated through to the middle of the twentieth century, beyond which time dawning awareness and concern about unintended consequences of land use, industrial emissions, over-harvesting and many other human activities for wildlife and the many societal needs met by ecosystems began to shift opinions and legislation.[19] This tide of awareness can be regarded with hindsight as 'disruptive science' challenging the primacy of the reductive paradigm.

Holism, humanity and ecosystems

The second paradigm of holism is concerned with the connections between disciplines and the need to think in multidisciplinary terms, based on the General Systems Theory.[20] The strength of holism is that it recognizes the existence and relevance of other components of the system, and of potential interactions between them, and that it therefore seeks to form balanced views between them as a basis for holistic action. However, the perceptions remain that disciplines are discrete and in practice often competitive. In 'real-world' decision-making, holism is often marked by the language of 'trade-offs' between the interests of different disciplinary views.

Holism brought with it a growing understanding of how all technologies and societal activities, formerly implicitly assumed to be independent, in fact interacted with and had wider consequences for other aspects of the environment and those dependent upon it. This drawing of connections can be traced in a great deal of early evocative writing warning of environmental degradation or catastrophe. This literature includes, for example, Harry Plunket Greene's 1924 *Where the Bright Waters Meet*,[21] Aldo Leopold's call for a new 'land ethic' in his 1949 *A Sand County Almanac*[22] (we will return to Leopold later in this book in Chapter 10) through to collation of shocking evidence of the adverse effects upon wildlife of persistent pesticides in Rachel Carson's seminal 1962 book *Silent Spring*.[23] This was backed up by the first sight of our small, seemingly vulnerable home planet viewed from space in photographs beamed back from the *Apollo 8* lunar mission in 1968, at a time of high-profile environmental disasters throughout the 1960s and 1970s, including the *Torrey Canyon* oil tanker spill in England, heavy metal contamination of marine ecosystems and the human food chain around Minamata Bay in Japan, and pollution from tons of buried chemical toxic waste beneath the Love Canal neighbourhood of Niagara Falls, New York, from 1976 to 1978. These formed part of a cultural sea-change with respect to humanity's conception of its relationship with the environment. In turn, they spurred responses such as instigation by the UN of the 'Man and Biosphere' programme in 1970 and the UN 1972 Stockholm Conference on the Environment, elevating public and political awareness of the global nature of the environmental problems unwittingly caused by humanity, their longer-term implications for ecosystem integrity and human development, and the need for concerted action to find solutions.

A number of significant advances in nature conservation can be traced to the early transition to the holistic paradigm between the late nineteenth and early twentieth centuries, as the broader ecological and human consequences of unconstrained development became better appreciated. Commonly, this focused on charismatic species as well as, as we have observed, those with sporting or other interests. For example, activities in the 1890s, eventually leading to establishment of what is now the UK's Royal Society for the Protection of Birds (RSPB), were initially driven by realization of the need to combat the near-extinction of the great crested grebe (*Podiceps cristatus*) to meet the demands of trade in 'grebe fur' (the bird's skin and soft under-pelt were popular as a fur substitute in ladies' fashions) and the head frill feathers of the adult bird. In the USA, the National Audubon Society was incorporated in 1905 with the mission '… to conserve and restore natural ecosystems, focusing on birds, other wildlife, and their habitats for the benefit of humanity and the earth's biological diversity'.[24] Other notable British special interest groups established in the first half of the twentieth century were the Pure Rivers Society, founded in 1926 to tackle the endemic pollution of fisheries and wider aquatic environments, and the Anglers' Cooperative Association (originally formed in 1948, subsequently re-formed as the Anglers' Conservation Association and, in 2009, as Fish Legal) to apply common law to take action against those responsible for pollution or other harm to fisheries. The key trigger for all of these initiatives was the drawing of linkages between consequences for wildlife and wildlife-related interests and the activities of other sectors of society.

As elaborated in Chapter 2, evolving legislation increasingly recognized the need to protect the environment from destructive human practices. The perception of flooding as an 'Act of God' came under increasing scrutiny, with the siting of development increasingly seen as contributory to its propensity to flood, together with growing awareness of the consequences of drainage schemes upstream contributing to more concentrated flood peaks downstream. The largely single-issue focus on land drainage that had dominated flood defence policy in the UK and across Europe up to and beyond the post-Second World War period also began to break down as connections were made with implications for declining habitat for wild birds and other species as well as the exacerbation of flooding in undefended land. This was significantly spurred by widespread flooding across England and Wales in 1947, followed by a tidal surge on the east coast of England in 1953 which caused widespread flooding claiming hundreds of lives. The 'Waverley Committee' set up to learn lessons from these events reported in 1954[25] with recommendations that subsequently significantly broadened UK attitudes to flood and coastal defence beyond land drainage. Requirements under the UK Water Act 1973 for new Water Authorities to 'have regard to the desirability of preserving natural beauty, flora and fauna …' demonstrated increasing understanding of interdisciplinary impacts. So too the building of fish passes into weirs and a growing acknowledgement and accommodation of wider concerns in flood defence schemes. Consequently, flood banks were 'softened' to provide some habitat for biodiversity, and there was a general shift away from harder bank reinforcement structures such as sheet-piling or concrete. Greater aesthetic and amenity considerations also began to influence the design of flood defences and other river engineering.

Statutory measures reflecting holistic appreciation of ecosystems

At the same time, agri-environment subsidy schemes began to re-circulate taxpayers' money to reward management or abandonment of land to achieve wider public benefits, significantly including biodiversity. However, it is a consequence of the perspective of holism that these subsidies and 'setaside' payments focused on only certain 'important' sites, implicitly failing to recognize and influence the increasingly damaging impacts of intensive farming across wider landscapes.

Holistic perspectives also began to expose the broader, formerly overlooked impacts of large dams around the world for ecosystems and human well-being. Constructed primarily with regard for the utility of stored water, hydroelectric power and other economically beneficial services, generally furthering the interests of those making or influencing dam-building decisions, there was growing recognition of broader consequences for hydrology and access to water resources lower in catchments as well as implications for sediment and nutrient flows, populations of fish, amphibians and other wildlife, including plant life and habitat-forming processes such as flood-plain sustenance and the natural regeneration of soils and grazing.[26] Although many political regimes around the world still tended, as some still tend, to overlook these consequences for less privileged sectors of society dependent upon the many benefits provided by rivers in a natural and functional state, the language of 'mitigation' grew more prevalent as a means to offset some of the increasingly widely recognized unintended impacts. Nevertheless, the primary design purpose of these schemes took precedence, with mitigation of wider impacts often a token, 'end-of-pipe' activity constituting a form of trade-off of the impacts of damaging activities on wildlife and people.

In the USA and Canada, across Europe and in various other developed nations, catchment management plans (CMPs) began to emerge in the late 1980s and throughout the 1990s as a more integrated form of environmental management at drainage basin scale. CMPs collated information and aspirations relating to different specialist management perspectives (generally including water quality, water resources, fisheries, flood management, recreation, conservation and navigation) for natural drainage basins, which had come to be perceived as logical, geographically defined management units. Although with hindsight sometimes criticized as collating different disciplines only to the extent of the staples binding the reports together,[27] these early CMPs nevertheless represented a significant step forward into holism from the parochialism of reductionism.

Under holism, budgets tended to remain 'ring-fenced' to address only specific tightly defined management disciplines. Modification of aspects of flood defence design for nature or heritage conservation, fishery access and protection or amenity and recreation therefore represented a net cost and risk to narrowly focused budgets, departmental objectives and engineering designs. Consequently, collaboration across disciplines was commonly regarded as a net cost and risk to the sponsoring budget-holder. These narrow, discipline-centred 'siloed' approaches and budgets gave rise to assumptions that there were always 'win–lose' trade-offs to be made in environmental management.

Segregation of people and nature in an age of holism

The conceptual separation of 'people' and 'nature' also persisted under holism, leading to a nature conservation model predicated upon preserving ecosystems in a 'natural' or 'wilderness' state. This approach served to protect many species and habitats that would otherwise have been driven extinct as well as preserving valued urban 'green spaces' and landscapes, and remains important in protecting the most endangered species and habitats in increasingly pressurized landscapes. However, the conceptual separation of 'natural' from implicitly 'exploitable' geographical areas came to define an era of 'fortress conservation' concentrated on fencing off, figuratively or literally, 'natural' designated sites or landscapes from potentially disruptive human interference. This diverted attention from the need to moderate practices outside of designated sites that degraded ecosystems.

At its extreme, this divisive thinking extended conservation management into various forms of environmental racism. Although taking an extreme form under the apartheid regime, including the expulsion and dispossession of black South Africans from their homelands and heritage during the construction of national parks[28] as part of a nature conservation strategy that neglected social needs,[29] South Africa is not unique. Concerns about the racial injustices have been raised in the establishment of national parks in countries as diverse as India[30] and Taiwan,[31] as well as the USA, where historic nature conservation and environmental concerns functioned politically as a coalition of groups with a variety of environmental interests, including outdoor recreation, wildlands, open space, public health and pollution, but which largely reflected the tastes of a white political and economic elite[32] and excluded racial minorities.[33] Indeed, the history of clearances and dispossession of Native Americans from land to be designated as national parks in the USA closely mirrors the historic creation of national parks in South Africa.[34]

The holistic paradigm nevertheless spawned various consultative processes recognizing that at least some other people also had rights. The nature of this consultation is now perceived as low on the 'ladder of participation'[35] (also known as the 'ladder of engagement'), effectively giving stakeholders little opportunity to participate beyond commenting upon courses of action largely predetermined by the decisions of a narrow cadre of experts and the already significant sunk costs associated with preferred design options.

However flawed and inflexible in practice, 'consultation' at least acknowledged that there were views beyond those of nominated experts and the governing elite, providing some capacity to modify management outcomes if not substantially to challenge basic design assumptions and decisions. This type of consultation is today increasingly referred to as DAD – decide-announce-defend – descriptive of 'expert'-led decision-making when predetermined proposals are simply announced to affected organizations and communities, with increasing amounts of effort then expended on defending the position taken.[36]

The breakdown of holism in a changing world

Many of the environmental 'shocks' of the latter half of the twentieth century constituted a form of 'revolutionary science' that came to substantially undermine the holistic paradigm. These included ecological and health crises stemming from bioaccumulation of pesticides intended for beneficial purposes, aerial and waterborne eutrophication arising as unanticipated consequences of land use and industrial processes, and stratospheric ozone depletion resulting from chemicals intended to improve human quality of life. Dramatic declines in farmland bird numbers across Europe, apparently largely in response to land use changes driven by the EU Common Agricultural Policy,[37] provided further evidence of the conceptual disconnection of nature conservation and farming across the landscape. Evidence of increasing harm resulting from species introduced beyond the natural ranges in which they had evolved, awareness of the vulnerability of birds and other animals to multiple human-induced stresses across their broad migratory ranges, and the pervasive impacts of climate change also pointed to wider linkages in the environment eluding contemporary fragmented management and resource use. This was accelerated further through additional 'shocks', such as extreme flooding events in the UK, including the intense 'millennium floods' of winter 1999/2000 and autumn 2000 and recurring severe flooding during the summer of 2007. These stimulated the government to commission the independent report *Learning to Live with Rivers*,[38] to evaluate the contribution of agricultural land to management of flood risk,[39] and to instigate the 'Pitt Review'[40] into the 2007 events.

These cross-disciplinary impacts eroded confidence in 'siloed' conceptions under holism. Further discoveries about the volatile and ever-changing nature of the climate, species distributions and their responses to changing land uses, diffuse pollution and other forces contributed to an increasing realization that there was no absolute baseline 'natural' environmental condition, be that post-glacial, at an arbitrary point of species recolonization thereafter, pre- or post-Agricultural Revolution, or pre- or post-Industrial Revolution. Land use practices became recognized as significant among the range of influences contributing to nature and heritage conservation, as well as the propensity of landscapes to detain or amplify flooding, and their ability to aid fish recruitment in rivers. This, in turn, revealed manifest shortcomings in spatial nature conservation designations, focused on preservation of habitat in a 'near-natural' state within a wider landscape implicitly surrendered for human exploitation. Indeed, the tight boundaries of reserved areas now appeared frequently to inhibit the succession processes responsible for maintaining habitat and species diversity or special interest.[41] Also, flood risk could clearly no longer be managed by hard defences with some retrospective accommodation for nature and amenity. Traditional flood defences were found often to present a greater risk to nature conservation than the floods they were designed to control,[42] while emissions from energy-intensive waste-water treatment technologies were recognized as ultimately posing more serious potential risks through climate change than the water pollution they were applied to abate. The impacts of people on catchments, through land uses, urbanization, infrastructure and the economic forces that drove them, were becoming more evident in terms of their contributions to flooding, biodiversity loss, fishery degradation, pollution, sedimentation, the costs of treatment of abstracted water, and a range of other factors with social and economic consequences.

Established conceptual frameworks and governance models were self-evidently becoming too blinkered to cope with growing awareness of the complexity of human–ecosystem interactions, exacerbated by the emerging implications of a changing climate.[43] Having battled in the reductive and holistic paradigms to erect the metaphorical or literal 'barbed wire' around designated nature reserves, there was a dawning of realization that the protection of ecosystems, wildlife and their vital functions and services ultimately depended upon re-engineering land use and other pressures in the wider landscape more sustainably to enable nature to move between and beyond designated sites, many of which were too small or too changed in character to sustain viable populations of organisms or wildlife communities of conservation concern.[44]

This awareness gave rise to concepts such as 'wildlife corridors' or 'ecological restoration zones' to allow organisms to move more freely to accommodate pressures, to repopulate non-designated areas and to respond to climate-driven changes in countryside that a former, more blinkered approach to nature conservation had perversely sanctioned, tacitly approving subsidy for ongoing and accelerating degradation through unsympathetic land uses.

New ways of managing ecosystems

Exceptions to this generality of fragmentation under holism included notable successes of some 'flagship species' conservation projects, such as India's tiger and elephant reserves and the giant panda in China, for which conservation of wider networks of habitats created a focal point for landscape-scale conservation. This step-change from a focus on local parcels of land to the integrated functions of entire landscapes, of which mobile and often predatory species may be an indicator, is one of the hallmarks of progress towards a systemic view of the interrelated nature of wildlife, humanity and our diverse activities, and the land and landscapes of which all are part.

The same approach to 'iconic' fish species requiring networks of connected habitats is beginning to influence river conservation.[45] It was also becoming realized that management of socially valued fish stocks focused on protecting or enhancing the ecosystems upon which self-regenerating stocks depend could simultaneously yield multiple additional societal benefits.[46]

This is all part of an increasing realization that nature and natural resource conservation initiatives can only become effective if meshed with economic systems advantageous to the livelihoods of local people, in other words dissolving the conceptual walls between disciplines and interests under holism. This landscape-scale connection between ecological functioning, economic activities and social needs was also expressed in the realization of the need for equitable recompense for land management to retain water-yielding landscapes for the advantage of often remote water-consuming beneficiaries across catchment scales. Nature conservation, fishery and broader environmental management would clearly have to be embedded not only within wider perspectives of ecosystem integrity and functioning but also the social and economic contexts interacting with them if sustainability were to be achieved. River catchments were therefore beginning to be understood as social as much as ecological constructs,[47] as were other ecosystem types with associated nature conservation values.[48]

Revolutionary science and decision-making

Beyond these changing perceptions about effective management of species, ecosystems, fisheries and other natural resources, faith in science and 'experts' to provide all the answers pertaining to wise decisions was also manifestly crumbling with a realization of the need to make management decisions in the face of complexity and increasing uncertainty.[49] Rather than allowing one narrow stratum of expertise and influence to guide decisions, there was a growing recognition that more robust and better-accepted and -understood decisions are more likely to arise through respect for the multiple, interdependent livelihood needs and value systems of all of the people that they inevitably affect.

These influences collectively began to frame a different conception of humanity's multidimensional interdependence with ecosystems, leading to development of the concept of 'post-normal science'[50] to address circumstances where '… facts are uncertain, values in dispute, stakes high and decisions urgent'. In fact, 'post-normal science' merely reflects the systemic reality that science, particularly reductive or holistic science, does not have all the answers, and that social perspectives and other forms of knowledge have equal relevance to decisions that have to be made in the face of incomplete knowledge.

This 'revolutionary science' acknowledgement of the need for stakeholder engagement, accommodation of different types of knowledge, and co-creation of options and decisions in equitable and sustainable decision-making breaks out of the confines of the holistic model. It marks a transition to systemic science that addresses entire socio-ecological systems to inform management and decision-making. Such paradigm shifts are, of course, rarely smooth, with the risk of entrenched self-interests creating 'high-level equilibrium traps', a concept first defined to explain why China never underwent an indigenous Industrial Revolution,[51] and now more widely applied to describe situations when there is no economic or other pressure to improve efficiency beyond a politically comfortable baseline. However, society's need to address the challenge of making an elective transition into a new paradigm is clearly evident in both the momentum for change as well as the likely dire consequences of failing to continue to make progress towards a new post-industrial model.

Stepping into the systems paradigm

The third paradigm of systems thinking, evolving in science since the 1960s and still emerging into environmental management, is concerned with understanding the properties of whole, integrated systems and the relationships of their components.[52] Crucially, it considers the 'emergent properties' of systems that cannot be deduced by analysis of their constituent parts, including, for example, the catalytic power of enzymes, the mental functions of interacting brain cells, emotional content conveyed by arrangements of musical notes, cellular functions performed by constituent chemicals and organelles, and the diverse ecosystems of this planet that interact to create a self-regulating biosphere.

Systems thinking recognizes that interdependencies and relationships between elements of systems are essential; without them emergent properties are lost. The natural world comprises nested systems at scales from the universal to the biosphere, the continent, the catchment, the human body, the workings of the cell, and subatomic particles. Relationships between humans and the ecosystems that support them constitute a fully integrated socio-ecological system. The implications of systems thinking for management of catchment systems have been studied in depth,[53] with the ecosystem services they produce defined as examples of emergent properties.[54]

The developed world is today grappling with the transition to systemic practice, with all the hesitancy and inertia that one expects in a period of paradigm shift. Systems thinking frames ecosystems as pivotal not merely to particular nature conservation, flood management or fishery interests but to the breadth of supportive services upon which environmental integrity and human health, security, economic opportunity and potential for fulfilment depend. Ecosystems are then managed not merely as exploitable resources or as net passive receptors of anthropogenic pressures or in protection of relic populations – an afterthought and a net constraint on 'development' as framed by reductive exploitation habits – but as irreplaceable sources of multiple beneficial services essential to all facets of human well-being.

The former separation of 'human' and 'natural' that characterized 'fortress conservation', 'flood defence' and 'exploitable resource' models then breaks down under systems thinking, which instead acknowledges the interdependence of ecosystems, their benefits to society, and the human populations that depend upon, exploit and influence them. This perspective is exemplified by the audit by the UN's Millennium Ecosystem Assessment (MA) of major global ecosystem types, all of them substantially degrading across the world under human pressure with a reciprocal grave prognosis for their capacity to sustain human well-being into the future.[55]

Systems thinking has proved particularly effective for integrating understanding of formerly disparate disciplinary perspectives and local activities with particular application to complex socio-ecological systems such as river catchments.[56] It recognizes both the validity but also the importance of integrating different perspectives such as the diverse inherent and utilitarian values of the natural world, the rights of all in society, including those as yet unborn, and of economic activities cognizant of these natural limits. Systems thinking has been helpful in stimulating early progress in the transition from localized and functionally blinkered conceptions of problems, and parochial solutions that may contribute to new problems at greater temporal and spatial scales and across other disciplines, towards understandings of catchments and other ecosystems as contiguous and internally interdependent socio-ecological systems. Progress towards systemic ways of understanding is today erratic and incomplete, and transition into policy and practice remains at best patchy. However, more systemic thinking is progressively emerging in society's interactions with the ecosystems that sustain it.

For example, reform of UK flood risk management policy under Defra's 2005 *Making Space for Water* strategy[57] confirmed a more holistic approach to flood and coastal flood management that '... takes account of sources of flooding, embedding flood and coastal risk management across a range of Government policies, and reflecting other relevant Government policies in the policies and operations of flood and coastal erosion risk management'. This was reinforced by the following Flood and Water Management Act 2010,[58] which also sets out an aspiration to consider '... natural processes'. Drainage, flood risk and water management policy in the UK has thus shifted significantly towards an increasingly risk-based approach that acknowledges the underlying natural processes behind flooding, increasing uncertainty in the face of a changing climate, and human involvement in both causes and consequences. Only the most severe floods are now routinely considered to be 'Acts of God', including such extreme events as Hurricane Katrina in 2005, which was the costliest natural disaster in US history, constituting one of the five deadliest hurricanes, in which 1,836 people lost their lives with total property damage estimated at $US81 billion.[59] However, even the apparent randomness of these events is becoming less certain as we better understand the likely contribution of climate change to extreme and 'unpredictable' weather conditions.

The pan-global Ramsar Convention of 1971,[60] concerned with protection of wetlands of international importance, ushered in a new era of systemic conservation of ecosystems. It was founded upon the 'wise use' of wetland resources through promotion of sympathetic and often traditional social and economic patterns that do not fundamentally erode the 'natural character' of ecosystems and associated biodiversity. This integration of ecological, economic and social strands within socio-ecological systems has since evolved to become one of the central tenets of sustainable development, which is itself an inherently systemic approach to human development. Growing recognition of the interdependence of these three strands in all habitat types and landscapes was reflected in the 1980 *World Conservation Strategy*,[61] documented as a global consensus and a defining feature of sustainable development in the 1987 UN document *Our Common Future*,[62] and committed to internationally at the 1992 Earth Summit in Rio de Janeiro. The inclusion of multiple human interests and disciplines constitutes not merely a matter of equity but also a reflection of the roles and interests of all stakeholders, including their shares in the costs, benefits, innovations and actions entailed in exploitation and management of ecosystems. Since the protection of nature is today a matter more of politics than ecology alone, with real consequences for the many and diverse beneficiaries of the services produced by ecosystems, social justice and biological conservation must go hand in hand if they are to flourish in the long term.[63] We are progressively realizing that retaining the vitality and diversity of ecosystems is central to continuing human well-being and must become a systemic consideration in the shaping of all aspects of human development activities.

Making the transition

In the 'real world', transition between paradigms is progressive, with the commonplace coexistence of more than one. All three of the dominant societal paradigms explored above in reality represent steps in a continuum of broadening awareness about the workings of the natural and human worlds and their inherent interdependencies. Each presents progressively deeper insights into how systems function as whole, integrated units, taking account of increasing interdependencies and supporting more sustainable management. There is, therefore, no 'right' and 'wrong' paradigm. Reductionism remains a powerful tool for deducing detailed knowledge, for example about how specific landscape functions take place, but this knowledge must then be applied in a systemic way if unintended consequences from management actions are to be averted.

We live today with increasing systemic understanding of the unbreakable interdependencies between ourselves, our economy and the supportive ecosystems of this planet. Yet our economy and its vested interests still remain stubbornly and reductively divorced from the ecosystems that ultimately support them,[64] while much of education remains focused on reductive disciplines without the necessary next step of connecting them into an integrated whole. Meanwhile, a great deal of decision-making remains top-down rather than involving a broad range of stakeholders potentially affected by environmental management outcomes, and mired in the flawed assumption that sustainability may be achieved by mitigating decisions made in one sector of human interest in holistic 'win–lose' trade-offs with other sectors. However, modern conceptions of flood risk management are clearly moving policy and practice progressively in the right direction,[65] as are emerging 'paying for ecosystem' markets (more on these later in this book) that connect ecosystems and their management to beneficiaries via markets.

Though we remain as yet well short of the full realization of systemic practices, society is clearly engaged in transition to systemic management of its relationship with the ecosystems that support the basic needs, economic activities and the conjoined future aspirations of all in society.

Practical management of markets and ecosystems is unlikely to change radically overnight, led by a new utopian multi-stakeholder reorganization of society. The precise form in which governance and economic systems will unfold as flood risk, biodiversity and agrobiodiversity, landscapes and other environmental assets are managed for optimal public benefit is uncertain. However, market mechanisms allied with evolving statute and common law remain powerful instruments to create incentives for the necessary innovation entailed in societal transition. Significantly, the transition has to be affected by reward for appropriate innovation, and hence the opening of opportunities rather than increasing constraints on creativity. Beyond legislators, regulators, legal practitioners, land managers and entrepreneurs, all sectors of society have a role to play in brokering this transition as all are active players rather than passive recipients in the socio-ecological system. The NGO sector, for example, has been instrumental in bringing new awareness and solutions to the attention of society, but must itself also evolve, making a transition from a focus on narrow disciplinary, 'single-issue' environmental and/or social perspectives through to broader ecosystem-based and stakeholder models.[66]

Vision and strategy are key factors here. If there is consensus about the sustainability principles to which society aspires then these can guide daily decisions on an opportunistic basis to make a concerted societal 'course correction' of values, markets, regulations, products, practices and land uses. Time is not necessarily the linear progression that it may appear; the future is not yet created but can be shaped by many potential 'bifurcation points' in the present.[67] For example, the Industrial Revolution model was based on a few innovations with distinct advantages which have since shaped global history, such as the liberation of energy from combustion of fossil carbon reserves. It is quite conceivable for radical ideas and powerful insights (such as sustainability) as well as novel products (such as 'clean' energy and manufacturing systems) and novel land uses (such as those taking place under novel flood risk management and 'paying for ecosystem services' schemes) to also constitute bifurcation points that open the human world to multiple new possibilities. It is therefore wholly possible to switch paradigms, developing in new directions based on novel sets of assumptions, such as those increasingly recognized as essential for the attainment of a sustainable future.

The considerable transitions and challenges necessary for concerted progress towards sustainability, equity and the kind of connected vision implied by the paradigm of systems thinking are many and diverse. However, this chapter has revealed some dimensions of the necessary transition. The next challenge is to accept them as essential steps to securing our long-term well-being and to work out how to put them into effect in the many complex relationships we have with land, landscapes, other vital ecosystem resources and each other. Further urgent 'mainstreaming' of this systemic approach will be essential to secure our ability to live fulfilled lives.

5 | A landscape at our service

Sustainable development is the defining challenge of our phase of human evolution, integrating environmental integrity and functioning, the needs of all people and the reformation of the economy to assure a future. Notwithstanding criticisms of the concept addressed in Chapter 2, the alternative is unsustainable development which will perpetuate the fragmentation and degradation of the past and overlook all we now understand about the intimate interconnection between ecosystems, our activities and those of all with whom we share a common biosphere. This can only serve progressively to undermine ecosystem integrity and the prospects of everyone. There is broad international governmental consensus about the urgency of sustainable development, notwithstanding the pace with which it is being addressed and diverging opinion about precise implications and priorities. The challenge of innovating ways of living that are equitable, fulfilling and in balance with the supportive capacities of stressed and degraded planetary ecosystems is truly daunting in scale, the more so for the need to achieve breakthroughs rapidly and consistently across strata of national and global society.

What is certain is that pragmatic and practical tools are needed to make sustainable development comprehensible and tractable in the 'real world', and to expose unsustainable decisions as short-sighted anachronisms. This wider perception enabled by supporting tools needs then progressively to internalize how, in our daily lives, all resources have a biophysical source and fate, and that all of our diverse activities have consequences both for the ecosystems with which we have co-evolved and for those with whom we share them. Without pragmatic tools that people can understand and use, putting sustainable development into practice remains tricky and contested, and will potentially be thwarted by established norms, vested interests and/or short-term exigencies.

The integrating science of ecosystem services

The term 'ecosystem services' covers various strands of resource management science emerging since the 1980s. 'Ecosystem services' are taken today to describe the many benefits derived by society from ecosystems. The often irreplaceable importance of many such services –

supply of fresh water, regulation of air quality and climate, pollination, culturally valued landscapes, educational and recreational opportunities, and so forth – has been substantially overlooked throughout human development. However, all provide the most fundamental resources for our economic activities and also underpin our health and potential to live fulfilled lives.[1] They are the ways we connect with the natural world and, whether recognized, overlooked or dismissed, the benefits that society derives from ecosystems pervade all aspects of our interests.

Many parallel strands of ecosystem services science were developed independently to inform management of different ecosystem types, such as wetlands, oceans, forestry, rangelands, sea shores and coral reefs. Despite inconsistencies in language between these schemes, it was formerly common to use the economic term 'goods' to describe tangible and tradable assets derived from ecosystems (fresh water, timber and fibre, fish, fruit, and so forth) and 'services' to address less tangible but nevertheless fundamentally important and economically valuable benefits flowing to society (such as soil formation and fertility, pollination, disease control, and air and water purification processes).[2] Notwithstanding the diversity of approaches and inconsistent terminology, the central and uniting power of 'ecosystem services' is that, by recognizing and increasingly quantifying and monetizing these benefits, many formerly overlooked uses and values of ecosystems could be brought into planning and other decision-making processes. If not valued, they are overlooked, and therefore effectively deemed worthless in decision-making processes. Consequently, degradation of the ecosystems that produce them as well as the benefits they provide for humanity is all but assured.

A unified classification of ecosystem services

The primary purposes of the UN Millennium Ecosystem Assessment (MA), initiated by former UN secretary-general Kofi Annan in 2001 as part of his report to the UN General Assembly *We the Peoples: The Role of the United Nations in the 21st Century*,[3] were to assess humanity's impact upon the natural world, the resulting implications for human progress, and options for future pathways of sustainable development.[4] To achieve this, the many disparate strands of ecosystem services science developed across different countries and habitat types were integrated into a single consistent classification scheme amalgamating the diverse beneficial 'goods' and 'services' within the umbrella term 'ecosystem services'. The new MA classification of ecosystem services enabled integration of pre-existing information consistently across global habitats and bioregions, providing the international community with a consistent means of recognizing the primary ecological resources that make life possible, profitable and fulfilling for human society. The MA starkly exposed the magnitude of threat that the global population faces owing to declining ecosystem functioning and productivity. The MA ecosystem service classification also provides a valuable framework to factor these often overlooked yet essential and irreplaceable benefits to humanity into future policy development and decision-making processes.

The MA classification categorizes ecosystem services into four qualitatively different sets: provisioning services; regulatory services; cultural services; and supporting services. Provisioning services are those tangible resources that can be extracted from ecosystems to support human needs, more or less synonymous with previous definitions of ecosystem 'goods' and including such assets as fresh water, food (both farmed and harvested), fibre and fuel, and so forth. Regulatory services include those processes that regulate the natural environment, including air quality, climate, water, erosion and pests. Cultural services include diverse aspects of aesthetic, spiritual, recreational, tourism and other values. Supporting services often lack direct economic utility but include processes essential to maintaining the integrity, resilience and functioning of ecosystems and hence the delivery of all other benefits, including, for example, soil formation, photosynthesis and water recycling. The full MA classification of ecosystem services is listed in Table 5.1, with definitions derived from the MA 2005 synthesis report *Ecosystems and Human Well-Being*.[5]

The MA typology of ecosystem services is neither perfect nor complete for all applications, but proved adequate for the needs of the UN Millennium Ecosystem Assessment in being broadly intercomparable across ecosystem types and climatic zones. It has therefore since achieved wide global consensus, supporting many studies and initiatives across the world. Local modification of this basic set of ecosystem services may be necessary for specific purposes. For example, the additional 'regulatory services' of *salinity regulation* and *fire control* have proved particularly useful in some arid countries,[6] and subdivisions of services such as 'disease regulation' can usefully be broken down into distinct costs of treatment and lost productivity where informative.[7] Nevertheless, the MA classification of ecosystem services reflects a useful and adaptable common framework reflecting the multiple generic interdependencies between people and ecosystems.

This modern take on our relationship with nature was not initially universally accepted by all scientists and environmental practitioners. For

TABLE 5.1 The MA classification of ecosystem services

PROVISIONING SERVICES

Fresh water
Food (e.g. crops, fruit, fish, etc.)
Fibre and fuel (e.g. timber, wool, etc.)
Genetic resources (used for crop/stock breeding and biotechnology)
Biochemicals, natural medicines, pharmaceuticals
Ornamental resources (e.g. shells, flowers, etc.)

REGULATORY SERVICES

Air quality regulation
Climate regulation (local temperature/precipitation, greenhouse gas
sequestration, etc.)
Water regulation (timing and scale of run-off, flooding, etc.)
Natural hazard regulation (i.e. storm protection)
Pest regulation
Disease regulation
Erosion regulation
Water purification and waste treatment
Pollination

CULTURAL SERVICES

Cultural heritage
Recreation and tourism
Aesthetic value
Spiritual and religious value
Inspiration of art, folklore, architecture, etc.
Social relations (e.g. fishing, grazing or cropping communities)

SUPPORTING SERVICES

Soil formation
Primary production
Nutrient cycling
Water recycling
Photosynthesis (production of atmospheric oxygen)
Provision of habitat

example, some in the conservation community perceived, as some still
do, that ascribing a utilitarian and potentially economic value to nature
does not reflect the importance of rare and characteristic species and
ecosystems. Furthermore, it imposes man-made values on that which
pre-existed humanity and economic systems. Aspects of this criticism are
fair; ecosystem services will not replace the importance of discrete public
policy and investment in all aspects of habitat and species conservation,
particularly for the most endangered examples of each. However, this
overlooks the intent of the MA classification of ecosystem services not
merely to recognize different forms of human utility but also to integ-

rate many different value systems across society. The 'inherent value' of nature constitutes both a cultural service (noting in passing that the decision as to which elements of nature are valued and which are not is an inherently anthropocentric choice) but also a supporting service reflecting its role in maintaining ecosystem integrity and functioning, for which it serves as an indicator. Recognizing this intent of integrating value systems, the perspectives of ecosystem services and nature conservation have vastly more commonalities than differences as, although we do not know the mechanisms by which many services are 'produced', most depend upon biodiversity and the integrity and functionality of ecosystems.[8] Nature conservation and ecosystem services can be expected to continue to converge – for example, as they are within the English government-commissioned 'Lawton Review'[9] into future management of sites designated of wildlife interest. The Lawton Review highlighted the need to think in terms of landscape-scale connections and processes, and to manage the nation's fragmented wildlife under the ecosystem approach if it is to remain viable and valuable to society.

The findings of the Millennium Ecosystem Assessment

When published in a series of reports between 2004 and 2005, significantly including the synthesis report *Ecosystems and Human Well-Being* in 2005,[10] the MA comprised the most authoritative assessment ever of the status of global ecosystems conducted by in excess of 1,300 scientists across 95 countries. It synthesized data and knowledge around the new ecosystem service classification scheme, demonstrating unambiguously the rapid decline in all the major habitat types of the Earth across all bioregions. It considered their capacity to support our industry and basic biophysical needs as well as human potential and equity into the future, illustrating in graphic and frequently shocking terms the 'overshoot' of resource exploitation by humanity beyond natural regeneration rates and the parlous implications of this for long-term human well-being and security.

'Well-being' itself was further broken down into the constituent categories of 'basic material for a good life', 'health, including feeling well and having a healthy physical environment', 'good social relations', and 'freedom of choice and action', particularly with respect to equity and fairness. Central to the MA approach is that people are integral to ecosystems, in dynamic interaction with all of their parts, recognized not only as beneficiaries but as a major driving force of ecosystem change which, in turn, has huge significance for human well-being.

The MA synthesis report concluded that around 60 per cent of global ecosystem services are being degraded or used unsustainably and that circumstantial evidence suggests an increasing likelihood of non-linear changes, such as the collapse of fishery or climate systems. Degrading ecosystem services were also demonstrated to have disproportionate impacts upon the poor, contributing to growing societal inequities and disparities. Among the MA's conclusions are that '... changes that have been made to ecosystems have contributed to substantial net gains in human well-being and economic development, but these gains have been achieved at growing costs in the form of the degradation of many ecosystem services, increased risks of nonlinear changes, and the exacerbation of poverty for some groups of people. These problems, unless addressed, will substantially diminish the benefits that future generations obtain from ecosystems.' Humanity's destructive patterns of development are severely compromising the integrity of the natural world and its capacities to support us into the future. The liquidation of the natural world is also threatening to derail the United Nations' Millennium Development Goals[11] to halve global poverty and hunger by 2015. This conflict between booming population and the declining supportive capacities of global ecosystems is one of the most profound challenges facing our collective future.

The MA has further eliminated blockages to action previously excused by uncertainties about the complexity and multiplicity of interactions between social and natural systems. These include knowledge gaps about how all ecosystem services are produced, the need for methods to monitor them, and the type of response necessary for humanity to develop in a more sustainable direction. The Millennium Ecosystem Assessment itself is a powerful demonstration of the value of this reclassification of ecosystem services for addressing focal questions, building consensus and spurring concerted action. As the synthesis report states,

> Everyone in the world depends completely on Earth's ecosystems and the services they provide, such as food, water, disease management, climate regulation, spiritual fulfillment, and aesthetic enjoyment. Over the past 50 years, humans have changed these ecosystems more rapidly and extensively than in any comparable period of time in human history, largely to meet rapidly growing demands for food, fresh water, timber, fiber, and fuel. This transformation of the planet has contributed to substantial net gains in human well-being and economic development. But not all regions and groups of people have benefited from this process – in fact, many have been harmed. Moreover, the full costs associated with these gains are only now becoming apparent.

The 'ecosystem approach'

Stemming from the scientific paradigm of ecosystem services is the management framework of the 'ecosystem approach'. The ecosystem approach is a structured strategy to found policies and management practice on the basis of the diverse services provided by ecosystems. It has been championed by the international Convention on Biological Diversity (CBD)[12] since 1995, the CBD itself stemming from the Rio de Janiero 'Earth Summit' in 1992. The CBD defines the ecosystems approach as '... a strategy for the integrated management of land, water and living resources that promotes conservation and sustainable use in an equitable way'.

The Convention promotes the ecosystem approach as the primary framework for action commended to governments across the world. It is supported by twelve principles and five points of operational guidance. These have been transposed into various operational processes by governments around the world. Through the ecosystem approach, ecosystem services provide a valuable and robust basis for more inclusively accounting for nature's supportive mechanisms. They therefore also provide a framework for determining, on a comprehensive basis, the implications – both positive and negative – for perturbations to ecosystems and their likely consequences for different sectors of society arising from development schemes, strategies and policies.

The power of ecosystem services

Ecosystem services recognize that ecosystems and their various processes are the fundamental capital underpinning commercial and social security and progress, providing a language with which to factor them into mainstream economic and policy decisions. This takes us forward from the Industrial Revolution paradigm in which the supportive capacities of the world were effectively perceived to be limitless and available for unconstrained exploitation by those who owned or annexed them. However, the contemporary market economy is not merely a product of the industrial age but also bears a considerable legacy from it which perpetuates the destructive habit of resource liquidation for rapid return of profit.

Since ecosystem services are unashamedly anthropocentric, defining the multiple benefits that people derive from the natural world, they are inherently amenable to economic valuation. They may therefore serve the important role of challenging or reforming markets and other assumptions about the relationship between humanity and supportive ecosystems. Ecosystem services represent an influential lever for more sustainable decision-making, protective of the ecosystems underpinning long-term human interests and well-being and reshaping the economy to that end.

Ecosystem services can therefore also help redress inequities in access to the benefits of ecosystems and natural resources. As Thomas Malthus observed in his famous work *An Essay on the Principle of Population*,[13] all previous generations had included a poor underclass created by an inherent lack of resources. Malthus ascribes this to continued population growth, potentially logarithmic as compared to what he saw as only a linear prospect for growth in food production. As revealed by the analysis in Chapter 1, the creation of a 'poor underclass' has as much to do with who controls basic resources as to absolute numbers. However, regardless of relative privilege, the gap between the advantaged and the disenfranchised will ultimately benefit no one as all ultimately struggle to meet their basic needs if climate systems and the many essential services of land and landscapes collapse or continue to degrade. Equitable sharing of ecosystem resources within their finite 'carrying capacity' is the basic prerequisite for a stable and sustainable future. Ecosystem services therefore provide a common language for understanding the interdependencies of diverse dependants and users of common resources, and for constructive and innovative dialogue between them. Each of the ecosystem services represents a different constituency of beneficiaries, who are potential 'winners' or 'losers' from decisions pertaining to the use or management of the environment. Making decisions on the basis of the connected set of ecosystem services can therefore make transparent in decision-making the wider ramifications for all in society of choices and actions that affect shared natural resources.

In ecosystem services, we then have a powerful tool for founding decisions not upon arbitrary opinion, the shifting sands of popular views or vested interests, but upon robust scientific principles stemming from the fundamental processes, habitats and organisms that sustain life. This can therefore inform debate about the equitable and sustainable sharing of natural resources, and also provide a 'lens' to reveal the reality and gravity of our current predicament and the urgent need for a change of direction. Indeed, the lexicon of ecosystem services seems ideally tailored to the pressing needs we have identified for a common and transparent method to collectively recognize and value, and negotiate about fair and sustainable use of, the many societal benefits flowing from land, landscapes, catchments and other critical shared ecosystems.

Seeing the systemic connections

Ecosystem services are one of a number of emerging conceptual frameworks based on the properties of systems, acknowledging their inherent complexities and interdependencies. However, to maintain the broader insights that they can embody, it is essential to apply the ecosystem approach as a fully systemic approach and not as a simple classification scheme. Viewed as a classification scheme, there is a real risk of slumping back into reductive or holistic ('trade-off') thinking and decision-taking, with their associated 'blind spots' and inequities. Applied systemically, ecosystem services instead embody a different world-view that leads to different awareness, analyses, technologies and outcomes. As Albert Einstein famously said, 'The significant problems we face today cannot be solved at the same level of thinking we were at when we created them.' Thinking from a systemic paradigm that acknowledges the inseparable interconnections between elements of ecosystems, facets of human resource use, anthropomorphic classifications of different environmental management disciplines and all sectors of society is vital for wise and enduring decision-making.

This is a significant step forward from many of society's inherited assumptions.

Misinterpretations and misapplications of systemic concepts and tools unfortunately remain commonplace today, as society grapples with the paradigm shift from reductionism or holism. There is a real and prevalent risk of systems approaches, such as ecosystem services, being seen as simple classification schemes and then applied on a reductive (maximize just one service) or holistic ('trade-off') basis rather than with full consideration of the interdependencies between components. We commonly see this in fragmented thinking about the systemic concept of sustainability, which can compromise achievement of sustainable outcomes. For example, most professionals may be able to recite the three principal strands of sustainable development – ecology, economy and society – but it is still all too common to see them misapplied as an arithmetic sum rather than as a fully integrated system, such as in stimulating the economy and social provision of employment as worthwhile benefits potentially to be 'traded off' against the surrender of habitat for industrial or residential development. In reality, this merely perpetuates the long-established economic model founded on capitalization of ecosystems and natural resources with unsustainable outcomes. Sometimes, of course, there are genuine trade-offs, such as where the management of land required for food production is modified to accommodate new priorities, such as supporting biodiversity and water yield, between which there may be some compromise. However, truly sustainable outcomes depend on going beyond mere acknowledgement of these three strands of interest, looking instead for innovations that allow each to be protected or enhanced simultaneously.

As ecosystem services increasingly influence national and international policy-making and practice, we have to watch for the same non-systemic myopia that has dogged the misapplication of many previous systems tools. It is helpful when introducing people into a new paradigm to acknowledge their expertise in addressing specialist sectors of the whole system – for example, in water quality or flood risk management – but it is important to demonstrate the benefits of applying this expertise in a broader systemic context that acknowledges interdependencies with all other connected management disciplines and their multiple beneficiaries. Sustainability is the net casualty of inadvertent or lazy 'sub-systems thinking', with the misapplication of systemic frameworks most often undermining the well-being of the most vulnerable in society while benefiting those already influential or with better-understood economic interests. It is only when we explore the full interactions of all ecosystem services simultaneously, not always in minute detail but certainly as an initial risk-based scan before undertaking more detailed analysis, that we can be confident of spotting consequences for all of the many beneficiaries of ecosystem services from all sectors of society.

The systemic reality is that human well-being and sustainable economic growth rest upon supportive ecosystems. This demands of us not 'trade-offs' but innovations to protect or enhance ecological and human well-being simultaneously as an interdependent whole. Sometimes, this may mean re-evaluating or ceasing exploitative habits that inherently undermine ecosystems and social well-being regardless of their profitability, but it can also liberate creativity and develop new markets with lower associated risks and higher security of return on investment in a world of shifting values.

'Mainstreaming' ecosystem services

The uptake of ecosystem services was significantly accelerated by the comprehensive review of the Millennium Ecosystem Assessment, with many nations and international institutions now internalizing the approach. It is useful to highlight just a few of the diverse, widespread and ever-increasing examples from around the planet.

There are many initiatives in the USA, including measures to reflect ecosystem services in forest land use and conversion,[14] an exploration of strategies for ecosystem-based management by the Great Barrier Reef

Marine Park Authority in Australia,[15] while the Natural Forest Conservation Programme and the Grain to Green Programme constitute two of China's foremost ecosystem services programmes.[16] More concerted attempts are being made in South Africa to internalize ecosystem service concerns into land- and water-use planning sectors.[17] In England, the 2010/11 restructuring of the English government's Department for Environment, Food and Rural Affairs (Defra) under the Natural Value Programme (NVP) is seeking to embed ecosystem services centrally into all of the department's activities and influences.[18] Globally, the UN- and EU-sponsored TEEB (The Economics of Ecosystems and Biodiversity[19]) programme has a more ambitious mission of factoring the planet's multi-trillion-dollar ecosystem services into policy-making, aiming to help save businesses, cities and regional authorities money while boosting the local economy, enhancing quality of life, securing livelihoods and generating employment. The TEEB programme's 2010 final report was preceded by a range of targeted guides for 'policy makers', business, citizens, and 'local and regional policy makers' to demystify and operationalize the approach in pragmatic and routine decision-making. Other global initiatives with broadly similar aims include the Natural Value Initiative,[20] which is business-facing, and the Natural Capital Initiative,[21] which aims to connect different sectors of society.

We are indeed in the early phases of a profound transition, with many serious attempts across the globe to bring the services provided by the natural world into the heart of governance. Land and landscapes have, and will continue to have, a central role in providing the many services upon which human interests depend, as indeed do the freshwater and marine, mountain, wetland and coastal fringe habitats with which they interact. They may therefore remain at our service provided human activities can be modified to take full account of their interactions with supportive ecosystems. Given the implications for long-established vested interests and anachronistic assumptions about the finite capacity of ecosystems and the rights of those who share them, this will be nothing short of a revolution.

6 | The great food challenge

The quest for adequate food has defined much of human progress. It imposes limits to survival and growth of communities and civilizations, and their capacity to evolve beyond basic biophysical survival needs. Pursuit of food sufficiency has spurred many innovations and novel technologies defining the nature and direction of evolution. Collaboration in the production, distribution and trade in food has significantly influenced our need to communicate and to develop economic and recording systems. The history of food production underwrites the history of civilization, and shortages are still a dominant factor hampering the development of environmentally stressed regions such as much of sub-Saharan Africa and the Indian subcontinent. Food, like water, is thus a resource that both connects and divides but always impels humanity.

Food production and distribution and the food-producing potential of land are also in profound transition. This chapter explores these changes using examples from around the globe but principally focusing on the UK, as this 'island nation' represents a microcosm of wider influences.

Privileged access to food

The people of the already-developed world are cosseted by many luxuries that we take largely for granted. Many in industrialized societies enjoy not only sufficient food but also a surfeit that leads to all manner of diseases of excess, including diabetes, obesity and heart diseases. In the USA around half of the food purchased is thrown away.[1] Profligate lifestyles mean that UK residents collectively throw away between a third and a half of the food produced on its farms, collectively accounting for an annual $20 billion and also 20 billion tons representing half of the import needs of Africa.[2]

Our profligacy is enabled not only by highly efficient modern farming systems, but also global trading systems enabling us to reap the considerable benefits of cheap food produced overseas. Our trading 'muscle' includes access to food produced by many less advantaged countries, where the imperative of earning foreign revenue can overwhelm the domestic priority of ensuring that everyone is adequately fed. Such is our privileged state that we may have lost sight of the umbilical link between our well-being and what we eat.

This ready access to foreign markets, and the consequent transition of our economies towards higher-margin activities, leads in turn to levels of food self-sufficiency in the major economies that are often surprisingly low and commonly declining. Definitive figures are hard to come by, but a 2009 report[3] suggested a decline in food self-sufficiency in Japan from 73 per cent in 1965 to 37 per cent in 1993, recovering to 40 per cent in 2007. According to figures from the UK government's Department for Environment, Food and Rural Affairs (Defra), the UK enjoyed just over 60 per cent self-sufficiency for 'all food' and a little under 73 per cent for 'indigenous type food' in 2008.[4] Strong domestic and international markets allied with a large land area mean that the US bucks this trend, with 128 per cent self-sufficiency in 2003. However, we must not confuse food self-sufficiency with food security, which also depends upon inputs of energy and chemicals, machinery and other infrastructure, and so can also be dependent upon imports and the maintenance of good relationships and strong trading links.

The UN Food and Agriculture Organization estimates that agriculture accounts for about 70 per cent of global water withdrawals.[5] Furthermore, since 1940, 80 per cent of Mexico's public expenditures in agriculture have been for irrigation projects. In China, Indonesia and Pakistan, irrigation absorbed more than half of agricultural investment and, in India, about 30 per cent of all public investment went into irrigation.[6] Food production thereby has a direct relationship with pressures on the world's already stressed water resources. It is surprising, then, that water remains a resource that we continue to take so much for granted, despite long-term recognition that it is a significant limiting factor globally to human well-being and development.

As living standards rise along with material expectations, concerns about water quantity and quality are set to increase. Between 1950 and 2000, global water use more than tripled. In some areas, especially cities, rapidly growing populations make demands on water far in excess of available supplies. Even when there is sufficient water, distribution infrastructure can be woefully inadequate, leading to inequities in access and further injustices in terms of who is supplied and who is not, with knock-on consequences for the ability of all sectors of society to produce or procure food. An estimated twenty-six countries with a combined population of more than 300 million people currently suffer from water scarcity, and projections suggest that sixty-six countries, comprising two-thirds of the world population, will face moderate to severe water scarcity by 2050. Water consumption is also skewed to the privileged; the average US resident uses 600 litres per day, the average UK resident uses 150 litres per day, of which more than 50 litres is used to flush toilets, compared to the total of 20 litres per day consumed by 1.8 billion people in developing countries who have access to a water source within one kilometre but not piped to their house or yard.

In fact, the water demands of the developed world are even more skewed as 'virtual water', comprising the water required to produce water-hungry food crops such as rice and wheat and materials such as cotton and wood, can result in the net diversion and export of substantial volumes of water entailed in their production from other uses in supplying, generally, developing nations. The 'virtual water' represented by goods imported into the UK to support our emancipated lifestyles represents in excess of 4,500 litres per person per day. These factors combine to keep the people of the developing world in hydrological, food, educational and economic poverty and to drive them downwards in a spiral of increasing incidence of disease, lost productivity and potential, and higher costs and time entailed in obtaining water and producing food.

The established world order does not have equity at its core with respect to access to food, water, productive landscapes and many other important resources. However, we would be foolhardy to assume that the world order to which we have grown accustomed is in any sense fixed.

The modern global challenge

If food shortages were prevalent throughout history, they are also very far from irrelevant to our future. Globally in 2005, 2.5 billion people in developing countries, representing nearly half of the economically active global population, relied on agriculture for their livelihood, while, by 2009, 75 per cent of the world's poor lived in rural areas close to the land that produced their food.

The world is also facing a future of food shortage as the global human population exceeded 6.7 billion in 2010 on its way to a projected ceiling of 9.5 to 10.5 billion by 2050. More mouths to feed, more feet to tread upon the Earth while, at the same time, not merely a slowing down of efficiencies in food production but a changing climate that compromises the productivity of current farming systems. While global food production will have to double from current levels by 2050 to keep up with the needs of a hungry world, climatic effects ranging from escalating temperatures to water scarcity and soil salinity as well as increasingly extreme and erratic conditions exposes vulnerabilities in our food production systems. Yields decline as the occurrence of crop failure rises. Research by the International Food Policy Research Institute[7] used world food prices are an indicator of the effects of climate change on agriculture, applying scenario analysis to determine that the impacts of climate change could double beef prices by 2050, while the prices of such staples as rice, maize and wheat could be escalated by as much as 37, 55 and 111 per cent relative to a future without climate change. Much of this change is related to climate change impacts depressing the production of food. Undoubtedly, global environmental change will affect food security not only through agriculture, if it does not adapt both to a changing climate and the need to curb its emissions, but also owing to the security of globalized supply chains, price variability and volatility, and the governance of food systems.[8]

At the same time, rising affluence and life expectations, particularly in the fastest-growing nations of the developing world, are resulting in a shift from a relatively lower-intensity cereal-based diet to a more resource-hungry animal-based Western diet. All of this exerts growing pressures on land, water and energy resources, with associated impacts arising from emissions to air, land and water. Significant issues of equity also arise from the polarized distribution of both higher-value food and its associated negative environmental impacts. This shifting diet, hiking the prices of food and other commodities sharply, saw the Chinese government reverting to its long-standing policy of food self-sufficiency after toying briefly with relaxing this to 90–95 per cent to allow imports to meet a growing shortfall resulting from spiralling domestic consumer demand for meat and dairy products.

The G8 major economies, meeting in July 2009, pledged to provide $22 billion over three years to increase international assistance for agricultural development. However, this is far from a long-term solution to the world's major food security issues, which will instead require systemic changes to our relationship with land and its multiple benefits across all nations. Nor does it adequately recognize that today's dominant economies are likely to face substantial challenges themselves.

The national challenge in England

In January 2010, the government launched a campaign to boost English food security, recognizing the magnitude of the challenges ahead. This twenty-year strategy, *Food 2030*,[9] was the first of its kind in the UK for over fifty years, setting out a vision of the food system in 2030 and some measures for its achievement. The report contains some visionary rhetoric, including that by 2030 'Consumers are informed, can choose, and afford, healthy, sustainable food. This demand is met by profitable, competitive, highly skilled and resilient farming, fishing and food businesses, supported by first class research and development ...'. The report emphasizes the importance of strong British agriculture and international trade links for food security, including the efficient reuse of wastes. Aspirations such as to 'Produce more food in ways that protect and enhance the natural environment' and to 'Invest in the skills and the knowledge that will help the industry prosper' are important in recognizing the need for food production methods that build upon rather than undermine ecosystems and communities. Although, in part, this intention might help redress the shortcomings of a long-running historic policy framework that substantially overlooked food production and the rural environment, there are elements within it that tacitly recognize that we in the British Isles, with our rich farming tradition and economic advantages, are far from exempt from the looming challenge of feeding our people.

As the human population of the UK rises towards 70 million, the role of farming and food production is coming into sharper profile. This is not only in recognition of the fact that it contributed £80 billion to the economy and employed 3.6 million people in 2010, but also that we are simply losing many of the historic privileges that had formerly allowed us to take food supply largely for granted.

The same principles and responses apply to virtually all already-developed nations. Not only do the global challenges of food production in a changing climate apply to all national interests, but we are living through a profound readjustment of economic power and its associated geopolitical ramifications. The world order with which we became familiar in the latter half of the twentieth century, born in the Cold War era defining the United States and its allies as the First World, the Soviet Union and its allies as the Second World, and the Third World comprising the remainder of non-aligned and neutral countries, has simply transformed into a new paradigm that goes far deeper than the dissolution of the perceived threat from a Soviet bloc. Rather, yesterday's superpowers are waning in influence, becoming progressively eclipsed by the rise of new major economic players, significantly including the BRIC nations: Brazil, Russia, India and China. These emerging economies not only comprise vast centres of population with massive internal markets, but also major and ever-strengthening global traders. By the end of 2009, for example, China had overtaken the USA as the world's largest market for vans and cars. In 2008, the EU was marginally ahead of the USA as the world's biggest economy but with China in third place and rapidly accelerating at a year-on-year growth rate equal to or in excess of 10 per cent.

This is far more than a simple geopolitical point. The competitive advantages that the former 'First World' enjoyed as a major economic player are in decline as the purchasing power of emerging economies out-competes for our formerly uncontested pick of the global market for cheap food, oil, minerals and other resources, and hence capacity to appropriate the produce of land, landscapes and farming communities from across the globe. Based on current trends, the price of food and other everyday commodities that we have taken for granted for two generations is bound to change, and to do so radically. Added to this is the fact that global trade is substantially fuelled by a fossil fuel resource that we now know to be in decline, competed for more aggressively by emerging economies, and which we must anyhow phase out if we are to stabilize the planet's climatic systems. Casual assumptions about secure and privileged access to food and other vital resources from elsewhere across the planet have to be seriously in doubt.

Mobilization for food production

In the late 1930s, Britain was importing over 55 million tons of food a year, the majority of this arriving from the USA and Canada but with sizeable quantities also coming from continental Europe and sometimes farther afield. The outbreak of the Second World War exposed the vulnerability of these supply chains as merchant vessels crossing the Atlantic became easy targets for the German navy. Furthermore, these same merchant ships were also required to transport troops, munitions and other supplies to war theatres. All of this emphasized the need for drastic reductions in the volume of imported food.

Britain mobilized in three significant ways. First, farming was modernized, and this effort was supported by establishment of the Women's Land Army to provide labour for the growing agricultural sector. Secondly, rationing was introduced, under which local shops and shopkeepers were provided with enough food for their customers, who were required to register with them and to purchase their food through a system of ration cards. The third mobilization was the 'Dig for Victory!' campaign.

Dig for Victory! was instigated in Britain at the outset of the war on a nationally promoted yet locally organized basis. Many formal gardens, parks, lawns, sports pitches and areas of unused land were converted into allotments of all sizes, with everybody on the 'home front' encouraged to grow fruit and vegetables as a contribution to the greater war effort. People were also encouraged to keep a few chickens or ducks for eggs and meat, and some communities established 'pig clubs' in which pigs were fed on kitchen scraps with the pork shared out when the animals were slaughtered. These diverse activities were backed up by education campaigns helping people grow food, fronted by posters in public places and backed up by a series of more detailed information leaflets detailing cultivation techniques.

Dig for Victory! was perceived as a great success, not only producing significant food but also mobilizing non-combatants around the war effort. Between the outbreak of war in 1939 and its cessation in 1945, imports of food were halved as the acreage of British land brought into food production increased by 80 per cent. It was estimated that over 1.4 million people had allotments by 1945 at a time when the British population was 47 million. However, the British government also realized even before the end of the war that people would continue to need to grow their own food well after hostilities ceased, as Europe settled down from major disruption and had to feed many starving people in formerly occupied lands. The campaign continued, providing people with home-grown food for many years after the war. Indeed, rationing continued until the end of 1954, with meat the last foodstuff to come off the ration list.

This period of austerity emphasized the value of the UK becoming self-sufficient in food, which became a focal point for government policy in the post-war era. This, in turn, led to the mass drainage and reclamation of land (as reviewed in Chapter 4), which we are only now beginning to reverse as we take account of the costs to broader ecosystem services and societal interests of this blinkered focus on land and landscapes as media for food production.

Cheap food at any cost

This shift in attitude towards food and the value of land in its production set in chain a series of changes in public perception, establishing an expectation of cheap food which continues to damage the environment and undermine long-term public well-being today. This was coincident with the rise of major supermarket enterprises that progressively began to displace local stores based on regional food chains. While this lowered food prices, making food cheaper for the masses, it also depressed farm gate prices and accelerated the intensification of farming practices maximizing productivity through methods largely blind to wider impacts on the environment, the many other ecosystem services provided by landscapes, and to rural communities. Thus, the inherently laudable intention of the government to encourage a trend towards accessible and cheap food after the Second World War had, by virtue of its reductive approach, the perverse consequence of consumers getting used to the idea that food should be as cheap as possible regardless of the wider and longer-term impacts of production on the environment and human well-being.

This set of events also created an impression that the sole function of land was the production of food and fibre, overlooking its many important roles in the water cycle, air purification, climate regulation, provision of traditional landscapes, sustaining biodiversity and many ecosystem services besides. The glut of food has not only made us fatter and more susceptible to the diseases of over-consumption, but has also resulted in the undermining of farmland job security, the indebtedness of many farms, particularly including smaller farm units, and the progressive impoverishment of the wider environment and its many broader, essential ecosystem services.

Intensification of food production has also been strongly implicated in a plethora of 'food scares'. For example, the perceived threat to human health arising from BSE, or 'mad cow disease', surfaced throughout Europe in the 1980s and seems to be attributed to feeding cattle with slaughterhouse offal from which prions (infectious agents composed primarily of protein) from sheep were able to invade and affect the brain matter of cattle, raising major concerns about onward transmission to humans. Other high-profile food scares in the UK have included concerns about salmonella contamination of eggs from the battery hen flock, and listeria contamination of milk and soft cheeses. The UK has also seen diseases of farm animals reportedly arising from intensification and increasing centralization of slaughterhouses, including devastating foot-and-mouth disease outbreaks in 1968 and 2001 as well as more frequent outbreaks of swine fever and fowl pest.

Questions remain as to the wider costs of food produced 'cheap at any cost', and its implications for sustainability. When we consider the broader implications for all ecosystem services, we can begin better to understand the ramifications for ecosystems and the longer-term interests of people, and so inform our concerns and their solutions. Even from a narrow business perspective, short-termism that damages its long-term supply base makes no sense.

A renewed focus on food security

Food self-sufficiency has varied dramatically in the UK over time, summarized in the government document *Food Security and the UK: An Evidence and Analysis Paper*.[10] For example, prior to 1750, the UK was around 100 per cent self-sufficient in temperate produce, declining to 90–100 per cent (except for poor harvest years) between 1750 and the 1830s. This slumped further to around 60 per cent in the 1870s, to around 40 per cent in 1914, and to a low of 30–40 per cent in the 1930s. Partly in response to the national policies addressed above, these figures rose to 40–50 per cent in the 1950s, escalating to 60–70 per cent in the 1980s. From this peak, food self-sufficiency declined slightly to 60 per cent in the 2000s. The self-sufficiency ratio of domestic food production to consumption declined between 2000 and 2005, and reforms of the EU Common Agricultural Policy together with trade liberalization pressures are expected to reduce domestic production in the UK and Europe further still. However, in the face of the potential for disruption to domestic food supply due to a changing climate, international energy concerns exacerbated by domestic fuel price protests, geopolitical tensions and international terrorism, there is a renewed focus on the issues of food security and self-sufficiency.

Planning for food sufficiency is not only a wise course, but is one that is essential in a world of accelerating change, not to mention as an indicator of economic competitiveness. (The value of food exports from the UK in 2008 stood at £13.2 billion, while we imported £31.6 billion worth of food in the same year.) The UN Food and Agriculture Organization's *Rome Declaration on Food Security* specifies as a moral and logical aspiration that 'All people, at all times, have access to sufficient, safe and nutritious food to meet dietary needs and food preferences for an active and healthy life'.[11] Yet we have to be mindful of the dangers of conflating a drive for sufficiency or security with a 'Dig for Victory!', which may inadvertently reinstate the perception of land and landscapes as little more than media for the production of crops and commodities. We now have enough knowledge about the production and value of a broader set of ecosystem services to inform wiser, innovative and more equitable land-use decisions and production practices.

Charting a wiser course

Two key principles emerge from this consideration of historic, current and necessary future uses of the land. The first of these is that the costs of all of our activities, particularly those as intimately connected with soil, landscapes and the flow of nutrients and energy as food production, are ultimately incurred by ecosystems. This includes not merely production but also the clean-up of wastes and the renewal of resources. Yet, beyond an ill-informed notion of some boundlessly renewable resource lying in the background, there is no serious conception of ecosystems and their inherently renewable yet finite functions from the majority of economic uses of land. It is therefore essential to consciously consider ramifications for all ecosystem services and their diverse beneficiaries in order to challenge inherited short-termism, informing sustainable and equitable decisions and planning for the long term, including addressing the rights and interests of generations as yet unborn. Scales of time and space and the natural limits of ecosystems are the domains of science, unlike electoral and corporate planning cycles, so it is essential to invest in and pay due regard to robust knowledge upon which to inform wise land-use decisions and to develop appropriate technologies.

The second principle flows naturally from the first. This relates to the dangers of too narrow a focus on food production alone, as indeed any other narrow preoccupation with just one or a few services. This virtually assures long-lasting collateral harm to the many other ecosystem services overlooked in decisions relating to the use of land and landscapes, and hence to their many beneficiaries. A narrow focus, by definition, merely perpetuates the outmoded 'exploitation economics' model of maximizing short-term benefits blind to their collateral impacts, which has led to the myriad problems facing the world today. Throughout this book, we have catalogued an unhappy legacy of land flooded as a result of degraded upstream hydrology, drained wetlands which displace biodiversity and remobilize sequestered carbon, and agricultural or urban land development practices that have historically degraded fisheries and the quality of life of local communities. These are just a few examples among the litany of narrowly conceived gains eroding many wider benefits.

Almost three-quarters of the land area of England and Wales is used for food production. The agricultural sector also has major strengths, including resilience over time. Various traditional land management practices have yielded environmental benefits including forming and maintaining treasured rural landscapes, managing wildlife habitats, providing a major sink for carbon, and operating in catchments vital for groundwater and surface water resources. Yet the UK food system also continues to exert substantial negative impacts on the environment, largely resulting from efficient responses to perverse economic drivers embedded in EU and UK agricultural policies. The 2007 economic study 'The total external environmental costs and benefits of agriculture in the UK'[12] emphasizes reasons for concern and for prompt and systemic action, valuing the positive annual contribution of agriculture in England and Wales at around £1.5 billion yet costing a narrow suite of its negative impacts on the environment at between £1 and £3 billion.

These negative impacts arise largely from soil degradation, water pollution, water resource depletion, greenhouse gas emissions (agriculture is responsible for 7 per cent of the UK's greenhouse gas emissions and the food supply chain as a whole is responsible for 18 per cent), biodiversity loss and localized flooding. This continued mining of important natural resources underpinning longer-term security and future well-being can clearly not continue indefinitely without serious consequences, affecting not only broader human interests but also food production itself.

England and Wales, the wider UK and Europe are not alone in this regard, with modern, input-intensive farming systems replicating these consequences wherever they constitute a dominant use of land elsewhere in the world.

Revolutions of many kinds

We are now rather better informed, not to mention equipped with tools such as our modern and fast-evolving understanding of ecosystem services, to avoid many of yesterday's unforeseen negative consequences. However, the challenges are many, multidimensional and touch upon all in society.

There is, for example, no question that we need a technological revolution to help us maximize food productivity while minimizing environmental harm. The Royal Society's 2009 report *Reaping the Benefits*,[13] for example, highlighted that 'Crop production methods will also have to sustain the environment, preserve natural resources and support livelihoods of farmers and rural populations around the world. There is a pressing need for the "sustainable intensification" of global agriculture in which yields are increased without adverse environmental impact and without the cultivation of more land.' The Royal Society report considered the importance of genetic modification and plant breeding as well as a diversity of other propagation technologies.

In conceiving of this technological revolution, we have to embrace more than the driving forces of the previous 'green revolution' of the post-Second World War era, which saw global food yields so dramatically elevated per unit area of land through the use of fertilizers and other scientific concepts. We have also to think today of a 'blue revolution', particularly in the drier regions of the Earth in which many developing nations lie, which also maximizes food production per unit of the critical limiting resource of water.[14] Beyond soil and water, we must extend our vision to invest in the science that will help us address all dimensions of an ecosystem-based approach to protect and enhance soil and crop management.

However, we have to look beyond the spheres of technology alone. As the Royal Society's *Reaping the Benefits* report also notes, technology solutions '... have failed to acknowledge that there is no technological panacea for the global challenge of sustainable and secure global food production'. Therefore, the report proceeds, 'Global agriculture demands a diversity of approaches, specific to crops, localities, cultures and circumstances. Such diversity demands that the breadth of relevant scientific enquiry is equally diverse, and that science needs to be combined with social, economic and political perspectives.'

We also have to respect and learn from the low-intensity technologies embodied in the multitude of traditional practices around the globe that have enabled people to produce food in often hostile landscapes. Examples include traditional wisdom applied to tap flows of water for cropping, seasonal grazing and irrigation in arid lands across the world, flood-retreating cultivation that exploits natural soil fertilization processes, polyculture systems that maximize productivity by mimicking the complex structure of natural ecosystems, post-medieval water meadow systems that harness warmth and nutrients from river systems to force grass growth during cold weather, and many more besides, literally from across the globe. Rather than ride roughshod over them, as the industrialization of agriculture often has on the back of a narrowly conceived model of development that places 'green revolution' techniques above those that are locally appropriate and part of the fabric of tribal and cultural life, we need to promote methods often developed over millennia that harness the natural flows of nutrients and energy across landscapes, or to be itinerant in making use of them. This may indeed be vital for the retention of agrobiodiversity, ensuring the genetic diversity of the plants and animals that we exploit to assure us of resilience to pests and changing climatic and other environmental conditions. Agrobiodiversity also provides a reservoir of genetic traits that may be useful in the future. Indigenous knowledge associated with traditional technologies may also make a valuable contribution to boosting global food security where medicinal and other applications as well as sustained productivity in specific local environmental conditions can be transferred between bioregions, or adapted into new techniques.

A wider and well-connected policy framework is a necessity, integrating people's needs and traditions, stimuli for commercial innovation and new markets for the private sector, and providing supporting regulatory and subsidy systems founded on the broad range of societal benefits and costs associated with land-use decisions.

In the developed world, the retail sector may be a key player. It exerts a profound influence on what farmers grow and how they do so, so it too has to be brought into this wider framework as a stimulus for change. Clearly, this cannot work for local regions or single nations alone, particularly under the increasing forces of globalization, so we must engage with the wider quest for transformation of land use for the production of food in the context of other ecosystem services at greater spatial scales such as pan-European and under the auspices of the UN.

Sacrifices or trade-offs?

Decisions pertaining to the allocation of land for different uses are fraught with difficulties not merely of a technical nature but also economic, as well as conflicting with vested interests and assumed rights. It is, for example, self-evident that different landscapes are best suited for different land uses and the delivery of specific benefits, such as the importance of some uplands for water capture and storage for the benefit of often remote water users living downstream. Yet ownership rights and agricultural subsidies may create greater economic incentives for uses beneficial to a small minority of people, conflicting with resource protection and public value for the many. Any food production system that purports to be sustainable has to tread this fine balance of meeting the needs of all in society equitably.

There is an established line of argument that addressing the growing food demands of a burgeoning human population is best served by intensification of existing farmland, accompanied by 'land-sparing' of other, as yet unconverted, habitats. This is akin to the US model which, in a big and relatively sparsely populated landscape, finds room for expansive, high-yield 'grain prairies', rangelands and feed lots but also vast tracts of wilderness. Not only is it possible thus to produce large amounts of food – as we have noted, the USA is 128 per cent food self-sufficient and farmed commodities form a lucrative export market – but the land-sparing of unfarmed land works well for vulnerable species of wildlife which require habitats within narrow environmental ranges. Elsewhere, it has been found, for example, that 88 per cent of Ghanaian bird species were likely to benefit from land-sparing of forests.[15]

Biodiversity 'hotspots', or land important for other key services such as flood plains for floodwater retention or upland wetlands for water capture and storage, can thus be protected or used predominantly for compatible purposes to safeguard their wider benefits to society. Furthermore, non-farmed land may be set aside explicitly for such important public services as water capture, as is seen in more crowded quarters of Australia, the USA and the UK, as well as for carbon storage (i.e. with incentives for conservation of remaining upland peatlands and tropical forests through payment mechanisms such as the UN-REDD programme[16]), amenity and safeguarding of aesthetically important regions (such as national parks and conservation areas), and more sustainable, ecosystem-based approaches to flood risk management.[17]

'Sustainable intensification', as identified in the Royal Society report, is a laudable goal, although defining what this means in practice is contentious. However, were it possible to increase the yield of crops in least developed and developing countries to those seen in developed nations using current advice and best practices, this could boost global food supply by 80 per cent and could do so without the need to bring new land into production, averting the requirement for land-clearing of unprotected areas with a loss of biodiversity and important ecosystem services.[18] Not only could this approach contribute to feeding the world by 2060, but it would eliminate potentially massively increased greenhouse gas emissions and the loss of countless species through conversion of undeveloped habitat, along with the ecosystem services they provide for the benefit of wider society.

However, intensification and land-sparing have associated risks. Their appropriateness to, and efficacy within, other countries have yet to be tested – for example, in smaller and more crowded European countries, where human populations, biodiversity and food production coexist rather than being largely spatially separate. In these situations, the approach may, for example, lead to an increasingly fragmented landscape, and also divert attention from the pursuit of less ecologically harmful food production methods. This, in turn, may perpetuate environmental externalities within the mass-market agribusiness industry, which also influences other countries less suited to such intensification of food production. The potential of disease transfer may be substantially increased when natural disease and pest control services are squeezed out by the intensification, representing just another risk that has to be addressed before rushing into a 'magic bullet' solution.

The counter-argument to that of intensification and land-sparing is that of de-intensification, which basically seeks a balance between food production methods and their impacts on ecosystems, ecosystem services and local communities. Organic farming is one such example of a clear intention to use land more wisely to produce food less intensively but nonetheless profitably, while simultaneously supporting enhanced wildlife, more natural hydrology and protected soil fertility and avoiding a build-up of pesticide and synthetic fertilizer residues. There are also many good examples of polyculture systems across the world that produce multiple crops simultaneously while also maintaining some important ecosystem functioning. A good example of this is the coffee plantations that occupy some 40 per cent of the land area of the Coorg uplands of Karnataka state in Deccan India, in which crop trees (used for timber, kapok, jackfruit and other food and commodities) provide shade and a water-retaining microclimate for the under-storey of coffee bushes as well as supporting climbing plants (including pepper vines and vanilla), and in which bird, butterfly and other life thrives. Some of these coffee estates are, however, heavy on water use, though others more remote from rivers and 'tank' storage rely on rainfall and soil moisture, albeit with lower coffee yields.

Between the extremes of the contained intensification of food production and lower-impact, multi-service approaches are a range of other intermediate techniques, including flood-retreating cash cropping, common across Africa, to serve food and market needs yet not permanently modifying wetland structure or function. And this exposes the fact that there are many types of food production, ranging from those that support subsistence needs, through to informal cash cropping and commercial farming at a range of scales. Part of the complexity, then, is that there is no simple 'either/or' decision. Rather, the virtues of different models of food production have to be balanced with protection of other important ecosystem services provided by land, set within local geographic and cultural contexts responsive to the needs of people rather than serving the global market.

Learning from the exemplars

If this sounds like a call for major innovation and a massive leap forward, that is because the urgency and currency of the challenge of sustainable food production and landscape management demands as much. However, within our history there are both ancient and more recent examples of food production systems that may contain important lessons for the future.

Water meadows, for example, were a British innovation in land use from the post-medieval period that tapped into natural flows of warmth, nutrients and moisture from groundwater-fed rivers and springs to boost grass production. Diversion of relatively warm river water during cold weather, guided across a 'herringbone' of channels along ridges and furrows engineered into the surface of water meadows, forced the early growth of grass, critically during the late winter and early spring 'hungry gap' when hay reserves were depleted and before the first flush of spring grass. It also enabled irrigation of the land with water bearing silt and dissolved nutrients, and could be controlled for weed management. The innovative water meadow system was extended further through the 'sheep-corn' system, with animals fed on the water meadows by day and then driven to the adjacent nutrient-poor downland hills by night, where their faeces and urine carried nutrients ultimately derived from flows in the river. Water meadow systems substantially boosted outputs from farmed land and the agricultural economy as a whole. So efficient were water meadows that they spread rapidly from the 1600s, becoming ubiquitous across the flood plains of most rivers of lowland southern Britain, and indeed in many other regions of Britain with appropriate river morphology and geology, through to their decline during the twentieth century. My 2005 book *Water Meadows*[19] provides not only a comprehensive overview of the history and operation of water meadows, but also case studies including those of estates where a massive up-lift in cereal production from hills fertilized by the sheep-corn system was far more economically significant than the production of hay and stock on the water meadows themselves. A handful of traditional water meadows remain operational today, some of which have been worked for in excess of four hundred years with never an input of artificial fertilizers or petrochemicals beyond the elevated nutrient loads now commonplace in our rivers and the energy used in routine meadow and channel maintenance. This is surely a glowing example of a food intensification technology that works with natural processes, enhancing some attributes of wildlife in associated channels and without imposing pollutant and energy use burdens on wider ecosystems.

Organic agriculture may be a more contentious instance to its opponents, but is a positive example to its proponents of an imperfect but pragmatic model of farming that balances the production of food and other farmed commodities with conservation of soil fertility, wildlife and water fluxes through landscapes, avoidance of use of persistent synthetic pesticides and commercial fertilizers, and respect for farming communities. Analogues of the organic model are found in certain other food accreditation schemes, but can also be observed in various 'back to the earth' movements, including an emphasis in some British towns and cities on food self-sufficiency which is seeing the resurgence of allotments. So too the emphasis on peri-urban cash crop and subsistence farming adjacent to developing-world cities as a means to bolster food sufficiency and economic opportunity, and particularly the empowerment of women. A similar principle lies behind the rise of farmers' markets across Europe, eliminating a great deal of the costs and carbon emissions associated with transport of produce in the modern retail infrastructure and helping people enjoy the regional and seasonal produce of their locality while improving farm gate prices for local producers. So too the 'Slow Food' movement,[20] established in 1989 as a non-profit, member-supported organization to counteract 'fast food' and fast life cultures driving the disappearance of local food traditions and people's interest in the food that they eat. The Slow Food movement campaigns for food that is of good quality and taste, produced in ways that respect the environment, human health and animal welfare, and where food producers are paid a fair wage. Slow Food has certainly tapped a rich seam of public concern as, according to the Slow Food UK website in December 2010, there were over 100,000 members in 150 countries dedicated to bringing about systemic change in the awareness and behaviour of communities around the world through projects that involve both producers and consumers. Reconnecting people with natural and nature-friendly processes of food production is indeed something of a *zeitgeist* for local communities, in direct opposition to the continuing influence and economic might of multinational agribusinesses. The locally adapted polyculture systems discussed previously represent further exemplars of highly productive yet wildlife-friendly food and commodity farming systems, highlighting yet again the value of traditional knowledge in maximizing benefits on an environmentally and culturally relevant basis.

Integrating approaches to food production

Food production, whether through direct cropping or else manipulation of ecosystems to maximize yield through diverse agricultural practices, is of course just one of the essential services provided to humanity by nature. Our historic trajectory of development has, as already discussed at length, emphasized optimization of individual or a few services for the immediate benefit and profitability of narrow and influential sectors of society while overlooking and thereby generally degrading other services, along with the opportunities of those dependent upon them. Yet the coexistence of food production with the other suite of services provided by nature is critical. Notwithstanding the drive of agribusiness, which has created a perception of farm productivity dependent upon inputs of commercial fertilizers, pesticides, patented seed and stock strains, drainage and irrigation technologies and other mechanical interventions, even intensive agriculture depends fundamentally upon the natural processes of soil formation and fertilization, water balance, predation of crop pests, natural pollinators, breakdown of pathogenic microorganisms, and the social infrastructure of farming communities and the traditional wisdoms they embody. Furthermore, as discussed in Chapter 2, farming systems that systematically undermine such public resources as the capacity of landscapes to produce reliable and buffered flows of clean water, diverse wildlife, regulation of the climate and the quality of air and water, and provision of treasured landscapes and leisure opportunities, can in no sense be considered either sustainable, ethical or, ultimately, economically rational.

While the diverse production practices outlined as exemplars above may of themselves be insufficient to address the wholesale transition of food production systems towards a more ecosystem-centred approach that respects and protects the many other essential services provided by landscapes, they do at least furnish us with key principles that can be brought into the mainstream. We can use these to guide potentially sustainable innovation in modern food production systems. For example, while we are not going to go back to the back-breaking manual control of flows upon which the post-medieval stewardship of water meadows was founded, we can still grade land using modern laser levelling and global positioning systems (GPS), backed up by microprocessor controls of sluices tied in with meteorological monitoring or predictions to better tap the natural flows of water, warmth and nutrients through landscapes in a sustainable way.

In summary, boosting production from degrading ecosystems to meet the spiralling demands of a burgeoning human population must not be met by intensification of a simple linear production model, optimized by a few technological tweaks. We have to learn from but move beyond the 'food per unit area' emphasis of the 'green revolution', take the best lessons from the 'food per unit of water' perspective of the emerging 'blue revolution' and of exemplar technologies from our past, and build upon them a more all-encompassing 'ecosystem revolution' approach to production of food, simultaneously with a range of other socially beneficial ecosystem services. This ecosystem revolution must account for, and invest proportionately in, the relationship of food production with the multiple functions of land and the people dependent upon them, thereby maximizing both societal value and long-term viability. Technology has a really important role to play, but this must be integrated with the sustainable harnessing of natural productive and renewable processes that simultaneously address traditional, ethical, ecological and economic and other broader contexts, including catchment water yield and flood regulation, nutrient cycling, support for biodiversity and a host of other benefits to society. Many traditional land-use practices have been responsive to the needs of local people and the renewable capacities of local ecosystems, and it is of overriding importance that we retain this wisdom and priority rather than be driven automatically towards the wholesale pursuit of intensification to serve global markets over and above local need.

Food for thought

We can no longer afford to think of food as a commodity to be taken for granted as we move into a future of changing geopolitics, climate and population. Neither can we think of land as solely a medium for food production. Nor can we rely on the legacies of empire and carbon-intense energy sources to continue to appropriate the world's resources on a privileged basis. We have to farm our way out of this jam, not through intensification of the existing flawed model but by innovation of more connected uses of land and the harnessing of natural flows of energy, nutrients and seasons, safeguarding or restoring critical ecosystem services. This is far more than a technical challenge. Innovation or the rediscovery of traditional knowledge will be crucial to the development of living landscapes that support fulfilling livelihoods and safeguard natural character. This will depend on reform of economic systems and incentives. Furthermore, in a globalized society, solutions also have to work at a global scale yet be scalable to local needs and conditions.

All of humanity has a stake in how we use land for the production of food and other ecosystem services, whether as victims or beneficiaries. Outcomes depend upon the decisions that we take and act upon now. For this, we require the insights of science, the innovation of business, the leadership of politics, the support of appropriate subsidies, the guidance of policies, and the livelihood aspirations of the people who actually produce and consume our food. Across society, there is a pressing need to better link the agricultural sector with environmental objectives, in which markets play a key role. As a 2009 report to the UK's Land Use Policy Group puts it, 'The scale of need is hardly surprising bearing in mind that most of the environmental services required by wider society (including the management of carbon, water, biodiversity and landscapes) are currently unrewarded by conventional markets.'[21]

In short, fragmented policy interventions or empty political rhetoric fall well short of the necessary transition towards 'living landscapes' that provide people with decent livelihoods in ways that simultaneously support our demands for food and the many other ecosystem services provided by the landscapes that support us. This will be essential to secure the long-term well-being of all who share the limited common ground of global resources.

7 | Valuing land and landscapes

Through the Millennium Ecosystem Assessment (MA) and a great deal of subsequent work, ecosystem services have vastly improved understanding of the intimate relationships between ecosystems and their many human beneficiaries at all scales from the local to the global. They have also made far more transparent the multiple economic and social consequences of permitting continued ecosystem degradation. Ecosystem services therefore offer a comprehensive and robust basis for articulating the multiple benefits provided by the functioning of ecosystems to all sectors of society, across social groups and cultures and across space and time. Since they relate to the ways in which people interact with the natural world, they also make the value of ecosystems comprehensible and tractable, and many of these values can be expressed in economic terms.

Where they addressed ecosystems, former traditional industrial-era markets were founded largely upon their potential for rapid capitalization rather than any inherent or non-market values, including their potential to support human well-being in the longer term.[1] Thereby, much of the worth of ecosystems was almost totally excluded from the workings of the market and so tended to be overlooked in decision-making. The history of economic valuation of ecosystems on a species- or habitat-specific basis has been fraught with difficulties, the perspectives of different sectors of society generally frustrating efforts to determine meaningful and consensual absolute values.[2] However, ecosystem services open up new potential to ascribe values to ecosystems in terms of their multiple services to humanity, whether currently used or of potential value to humanity now and for future generations. This in turn supports the progressive internalization of the value of ecosystems and their many beneficial processes in the economy.

Counting ecosystem benefits to society

Although much of the land of the UK, Europe, the USA, South Africa and many other nations is in private ownership, many of the ecosystem services produced by landscapes support public benefits occurring well beyond the landholding. These benefits are realized by

TABLE 7.1 Benefits potentially derived from ecosystem services, with an indication of their scale of impact

Ecosystem service	Geographic and temporal scales of benefit (and illustrative set of potential beneficiaries)
PROVISIONING SERVICES	
Fresh water	Catchment scale (public supply, water and water-using industry and irrigation)
Food (e.g. crops, fruit, fish, etc.)	Local scale, seasonal (private intensive crop production and low-intensity cropping in either public or private hands)
Fibre and fuel (e.g. timber, wool, etc.)	Local scale, seasonal (private or public harvesting, grazing, building materials, etc.)
Genetic resources (used for crop/stock breeding and biotechnology)	Potential wide-scale and long-lasting benefits (private or public harvesting or breeding of genetic resources)
Biochemicals, natural medicines, pharmaceuticals	Potential wide-scale and long-lasting benefits (private or public harvesting or breeding of resources)
Ornamental resources (e.g. shells, flowers, etc.)	Local, but potentially long-lasting (private or public harvesting)
REGULATORY SERVICES	
Air quality regulation	Local to medium-range (public benefits to air quality and health)
Climate regulation (local temperature/ precipitation, GHG sequestration, etc.)	Local and global benefits, long-lasting (public benefits arising from climate stability)
Water regulation (timing and scale of run-off, flooding, etc.)	Catchment scale, long-lasting (public benefits from stable flows and flood attenuation)
Natural hazard regulation (i.e. storm protection)	Localized beneficiaries, long-lasting (public benefits from mitigating extreme events)
Pest regulation	Localized and enduring (public and private benefits from natural pest regulation)
Disease regulation	Local or medium-range and enduring (public and private benefits from natural disease regulation)
Erosion regulation	Catchment-scale impacts that may be long-lasting (public and private benefits of soil conservation and reduced siltation of waterways contributing to declining biodiversity)
Water purification and waste treatment	Catchment-scale impacts that may be long-lasting (public benefits through absorption of waste materials and improved quality of water supplies)
Pollination	Localized and enduring (public and private benefits from natural pollination services)

CULTURAL SERVICES

Cultural heritage	Localized or medium-range and enduring (public benefits from maintaining culturally important sites)
Recreation and tourism	Localized to medium-range and enduring (public benefits from amenity and private benefits from profit from tourism and recreation activities)
Aesthetic value	Local, medium-range and potentially global, enduring (public and private benefits provided by landscape)
Spiritual and religious value	Local, medium-range and potentially global, enduring (public benefits supported by landscape functions)
Inspiration of art, folklore, architecture, etc.	Local, medium-range and potentially global, enduring (public benefits supported by landscape functions)
Social relations (e.g. fishing, grazing or cropping communities)	Local to medium-range, enduring (public benefits through habitat support of local communities)

SUPPORTING SERVICES

Soil formation	Catchment scale, long-lasting (public benefits through creation of fertile soil)
Primary production	Local, medium-range and potentially global, enduring (public benefits through productivity of ecosystems)
Nutrient cycling	Catchment scale, long-lasting (public and private benefits through maintenance of productive cycles, fertilizing soils and metabolizing potential pollutant nutrients)
Water recycling	Catchment scale, long-lasting (public and private benefits through renewal of water systems)
Photosynthesis (production of atmospheric oxygen)	Local scale, long-lasting (public benefits through oxygen production and carbon dioxide sequestration)
Provision of habitat	Local, medium-range and potentially global, enduring (public and private benefits through maintenance of characteristic biodiversity)

Source: Modified from Everard and Appleby (2009)[4]

many constituencies at spatial scales, ranging from agricultural benefits from soil formation at the local scale, production of fresh water and regulation of flooding at catchment scale, contributions to air quality regulation with regional benefits, and the sequestration of climate-change gases with global impact. It is essential to take account of these extended values in decision-making if they are not to continue to fall victim to narrow commercial interests.

These publicly enjoyed rights effectively constitute a set of 'commons', nested at different spatial scales. They also implicitly take account of temporal scales from immediate commercial potential to the safeguarding of the integrity of productive ecosystems themselves and their potential to enable future generations to meet their needs and aspirations, consistent with the spirit of the 'Brundtland' definition of sustainable development: 'A form of development that meets the needs of the present without compromising the ability of future generations to meet their own needs'.[3] One of the consequences of this broad span of spatial and temporal scales and of cross-disciplinary and multi-stakeholder perspectives is that ecosystem services can better inform sustainable decision-making, but only if they are valued and therefore weighed in public and private decision-making. This inherently challenges established concepts of property and rights handed down from less enlightened eras of law-making, accounting and privileged ownership. Nevertheless, without taking account of the breadth of benefits and costs arising from human activities on ecosystems and other people, development will not proceed on a sustainable footing.

It is possible to generalize about the predominant beneficiaries of each ecosystem service in the MA set, also ascribing to each the range of geographic scales from the local to the global at which benefits are most likely to accrue. One such published breakdown is reproduced in modified form in Table 7.1.

On the basis of these benefits and beneficiaries, it is then possible to attempt to monetize the ecosystem services affected by management decisions. However, to understand why this is useful, we have first to consider the meaning of monetary values themselves.

Keeping the house in order

For something that we handle virtually every day, money is notoriously tricky to define. Turning to the dictionary, we are presented with various superficial descriptors of coinage and paper tokens before then 'drilling down' into conceptual strata that typically speak of money as a medium for exchange rather than something of intrinsic worth. Likewise the economy, a perennial obsession of our politicians, is most commonly defined in dictionaries as relating to the concerns and resources of the community and to governance systems geared towards the prudent stewardship of these resources. The economy, then, together with its many associated institutions, is merely a means to keep our house in order with respect to the stewardship and exchange of resources.

It is nevertheless clear from considering ecosystem services that, for all the sophistication and complexity of the modern market – trading in futures, derivatives and other increasingly virtual concepts – the underlying economy has tangible, biophysical roots ultimately reaching back into essential commodities such as fresh water, food and fibre and other ecosystem products. For all the complexity of the modern market, it is biodiversity and the ecosystems of this planet which ultimately underpin the economy, and ecological dependencies can ultimately still be traced in all types of business.[5] We have also seen how it is ultimately ecosystems which underpin other aspects of human security and progress. This is reflected in the emphasis of most international aid programmes supporting developing world economies, which now focus on rebuilding the supportive capacities of degraded ecosystems along with local human capacity to manage them on a more sustainable basis.[6]

Economy and ecology, then, are far more intimately interconnected than might be assumed from the often bizarre behaviour of the market. Indeed, the very root of the word 'economy' is the Greek *oikos*, meaning 'house', which is a common root to the word 'ecology', which is concerned with the interactions of organisms with each other and their surroundings.

This semantic preoccupation is of far more than academic interest, providing a foundation for understanding a system as old as civilization itself. Innovation of technologies to divert various of the ecosystem services generated by living landscapes for human advantage were the basis of evolving patterns of agriculture which, in turn, freed humanity from the opportunism and uncertainties of hunting and gathering. This step-change paved the way for settled lifestyles and further differentiation of responsibilities within communities, marking the dawn of civilization. Money, it is widely believed, was one of civilization's earliest inventions to facilitate the exchange of human and natural resources across increasingly complex societies, supporting the development of social structure, ventures, edifices and enterprises.

We are now increasingly recognizing what a dominant player humanity is in the ecology of this planet, and hence the powerful role of the economy in both environmental problems and their solutions.[7]

Markets for ecosystem services

There are established markets for some ecosystem services. Food, water, fibre and some other 'provisioning services' have been traded for millennia, albeit often externalizing various environmental and social costs entailed in their production. Monetization of these services is therefore relatively straightforward as they are already internal to the economy.

For other services, there are emerging markets. For example, the notion of carbon trading as a means to address the 'regulatory service' of climate regulation would have been laughable as short a time ago as the mid-1990s. However, today there are national schemes for carbon trading around the world, in addition to regional mechanisms such as within the EU, and international markets too.[8] These include, for example, the UK's Climate Change Levy and ongoing congressional debate in the USA about developing a cap-and-trade system for greenhouse gases building upon an existing national market to reduce acid rain and some regional markets in nitrogen oxides, international market instruments including the EU Emissions Trading Scheme, and commitments by national governments and international institutions to the further use of market instruments to decarbonize the economy. Notwithstanding the imperfections in these emerging markets in climate stability, they do at least begin to internalize this important service in the mainstream of the economy.

Arguably, agri-environment subsidies can also be seen as imperfect emerging markets. This is because, in the absence of true markets for many crucial ecosystem services, subsidy systems can act as near-market instruments to recirculate public investment, largely gathered through taxation, back into land management for public benefits. We have seen progress towards this at the European scale under the 2003 reform of the Common Agricultural Policy (CAP), which explicitly swung the emphasis of subsidy from market price support ('Pillar 1') towards rural development and environmental measures ('Pillar 2'), underlining shifts in policy interventions from cementing private gain towards promoting public benefits. However, its practical implementation in the UK's Environmental Stewardship scheme and indeed elsewhere across Europe demonstrates a conceptual shortfall in what precisely constitutes socially beneficial uses of land.[9] As we have already seen, the US system works somewhat differently with a general presumption towards 'land-sparing' to support a variety of beneficial human uses of landscapes taken out of production, including support for leisure activities, water production, biodiversity and 'wilderness'. Globally, the

network of Biosphere Reserves, World Heritage Sites and other protected sites reflects a similar safeguarding of places that the public
value for cultural and ecological reasons, conferring in them inherent
values superseding short-term economic gains that could be made from
their commercial exploitation. This is merely a different approach to
subsidizing landscapes of value for a generally poorly defined basket of
wildlife, water-yielding, amenity and other ecosystem services.

Other markets are completely absent, including, for example, for
services such as nutrient cycling, soil formation, pollination and
catchment-scale hydrology, though various strands of legislation and
policy (such as England's *Safeguarding Our Soils* strategy[10]) may recognize
aspects of their public value.

This inconsistent internalization of ecosystem services in a market,
the principles of which are largely built on capitalization rather than
protection of ecosystems, presents difficulties for the consistent valuation of all ecosystem services.

Valuing ecosystem services

Valuation of ecosystems and of ecosystem services is not a new preoccupation. It was substantially accelerated by the work of Partha Sarathi
Dasgupta throughout the 1970s on the economic theory of depletable
resources, green accounts, and the importance of essential ecological
services and environmental resources for poor countries.[11] The field of
environmental economics flourished thereafter, particularly following
the 1992 'Earth Summit' on sustainable development at Rio de Janeiro.
Many methods have subsequently been developed to monetize the functions of nature in terms of the goods and services exploited by humanity.

It is progressing today under various international initiatives significantly including The Economics of Ecosystems and Biodiversity (TEEB[12]).
TEEB was set up with the explicit aim of 'mainstreaming' the economics
of biodiversity, overcoming the invisibility of many essential services
that nature provides to underpin the sustenance, economic interests
and opportunity of humanity. TEEB states that 'If these costs are not
imputed, then policy would be misguided and society would be worse
off due to misallocation of resources'. A 2010 TEEB report advises that
the case for saving species is '... more powerful than climate change'.

Before these developments, there had already been many attempts
to apply economic valuation to 'nature'. These have taken a diversity
of approaches, often with some subjectivity or taking the form of contentious survey-based 'willingness to pay' or other methods to ascribe
values to species and/or habitats of particular interest, or of landscapes

or the amenities they supply. The results were variable in quality and usefulness, seeming to depend substantially upon the people polled and the methods applied, and therefore lacking in objective reality.

Ecosystem services have provided a more rigorous and economically relevant framework against which to consider the value provided by ecosystems to society. Framed as they are around benefits to people, ecosystem services are therefore inherently amenable to economic valuation, notwithstanding some practical difficulties. They cover the diverse perspectives and value systems of different beneficiaries around the four categories of provisioning, regulatory, cultural and supporting services, thereby providing a broad perspective on the societal worth of ecosystems for current and future generations.

An important milestone was set in the history of environmental economics with publication of a paper by Bob Costanza and colleagues in 1997, entitled 'The value of the world's ecosystem and natural capital'.[13] Costanza et al. conservatively estimated the value of all of the Earth's ecosystem services at $33 trillion a year, on the basis of replacement costs at current market rates. This was close to the world's total gross domestic product at the time. Costanza's paper still remains the best-known, albeit speculative, attempt to ascribe monetary values to the ecosystem services from which society benefits on a global scale. One of the conceptual troubles with this conservative and widely cited 'ball-park' estimate of nature's collective worth is obviously that it seeks to place monetary values on essentially non-monetary resources, for which substitution may not anyhow be a viable option. So the '$33 trillion a year' figure has its critics, with even Costanza et al. emphasizing that it is merely indicative and without absolute meaning, yet even if the true worth were $3.3 trillion or even $0.33 trillion a year, the figure is hardly insignificant!

However, we have to understand here that the primary purpose of economic valuation is not to boil everything down to a tradable figure. Rather, it is to indicate the relative strength of different value systems using a common 'currency'. Costanza et al. were explicit that their now famous $33 trillion annual figure had no objective reality. Rather than commoditizing nature, the primary purpose of valuing ecosystem services is to bring different value systems (including the various 'non-use' values of which the ever-elusive 'inherent value' of nature is part) into the decision-making mainstream. A decade and more on, we now have the benefits of far greater consensus about the ecosystems approach and the support of the unified ecosystem services classification developed by the Millennium Ecosystem Assessment. All of this adds rigour and acceptability to economic valuation, for which suitable valuation methods can then be applied or developed.

As we have seen, some provisioning ecosystem services have well-established albeit imperfect markets, others such as climate change have emerging markets, while many services are entirely external to the market. Values for some currently externalized ecosystem services can be captured from analysis of the compensatory or direct costs of their loss or degradation, such as soil erosion or nutrient enrichment, impacts on climate change and loss of natural pollinators and crop disease predators, or the travel, equipment and accommodation costs necessary to gain access to recreational fishery value. Valuation of other ecosystem services currently remains more complex for a range of practical reasons.[14] These may include where there is insufficient scientific knowledge to quantify ecosystem services of potentially significant value, such as the contribution of riparian reedbed habitat to regulation of air quality through processes such as physical settling of particulates and chemical breakdown of pollutant gases, or its various roles in regulating microclimate.[15]

'Supporting services', which primarily relate to internal processes that keep ecosystems intact and functional, and from which all other services flow, are notoriously difficult to value. Some do have near-surrogate market values, such as the costs of management agreements to maintain habitat identified by society as of nature conservation value. This value is, however, dependent upon shifting tides of public opinion and policy priorities. Other services are far more difficult to quantify, with 'cultural services' in particular sometimes subject to considerable cultural relativism. For many ecosystem services, we have to seek surrogate markets of one form or another, such as the 'expressed preferences' of people polled by those notoriously fickle 'willingness to pay' surveys, or more reliable 'revealed preferences', such as actual annual spend by visitors to travel to and enjoy particular nature conservation or other recreational sites multiplied by actual or projected visitor numbers.

Measures such as 'sense of place' or 'heritage value' are even less easy to calculate. Some of these less tangible, aesthetic aspects have been tested under case law, reflected, for example, in the quanta of damages awarded to recognize the 'amenity uplift' of the contribution of landscape and environmental quality to angling experiences distinct from the capture of fish.[16] Other environmental properties are hard to quantify in meaningful consensual terms, such as the inspiration of art, folklore or spiritual value, all of which are subject to substantial cultural relativism. However, as one example, for societies revering ancestors and the landscapes and landforms they are believed to inhabit (such as African Zulus, Australian Aboriginals or Native Americans) they are beyond monetary value. Other services defy most rational economic assessment methods, such as the 'glue' values associated with ecosystem integrity, resilience to environmental 'shocks' and maintenance of options for future development.[17] This is not, however, to say that they are not of immense value, so consideration of their likely significance must be weighed in decision-making even if quantification is not possible.

Methods for valuing ecosystem services

Guides are available[18] to the many methods that can be applied by economists to value ecosystem services, although it is important to observe three important cautionary notes. First, we have to avoid 'double-counting', for example by ascribing the same basis of calculation to the provisioning service of 'fresh water' and the supporting service of 'water recycling'. Each is a different ecosystem process, which implies a different set of benefits, beneficiaries and associated market values; it is essential not to replicate the same valuation, thereby skewing the net benefit calculation. Secondly, we have to remember that ecosystem services describe a whole, integrated system, and unsafe outcomes are likely to ensue if we select only a few services of interest and implicitly discard the rest. In so doing, we risk valuing and managing for desired outcomes and discounting broader ramifications, which is precisely how we created today's legacy of environmental and societal problems in the first place before systems approaches gave us the power to think in more integrated ways. Thirdly, it may in practice be impossible to ascribe reliable values to all services. For example, while we know that salt marshes constitute crucial breeding and nursery habitats for marine fish species of commercial and recreational value, potentially also with cultural importance as well as playing key roles in maintaining the broader integrity of marine ecosystems, we may lack the practical tools to relate the contribution to fish stocks back to changes in the state or extent of this habitat.[19] In this case, it is important not to suggest unrealistic or unsubstantiated values, but to weigh in decision-making that this benefit, though not monetized, is considered likely to be significant by experts. The Precautionary Principle should be invoked in such decisions, effectively to take account of the likelihood of positive or negative impacts, so a nominal value becomes implicit in the final decision rather than the benefit being wholly overlooked, as remains too commonly the case today. We can use identical methods to account for both benefits and costs, or 'negative values'.

In any assessment, where it is not possible to quantify and monetize ecosystem services, this should be stated alongside economic values deduced for other services, perhaps supported by expert judgements of the likelihood of scale of impact, to ensure that these important services are not simply omitted from decision-making.[20] All these practical issues have been applied in full in ecosystem-service-based assessments of a river catchment and an estuarine 'management alignment' scheme in England,[21] river habitat restoration,[22] supporting regional government decisions about land-use changes in the east of England,[23] and also applied in part in numerous other management scheme appraisals, particularly for flood risk management.[24]

Regardless of practical difficulties with valuation, the economic model implicit in the systems-based world is that the production of ecosystem-derived benefits (including positive management to protect or restore them) has tangible economic value, 'consumption' of these services should attract proportionate charges, and actual or theoretical markets should connect production and consumption of services bounded by the 'carrying capacity' of ecosystems. The comprehensive systems framework of ecosystem services, the high degree of global consensus around the MA classification, and their inherent compatibility with economic valuation techniques equip ecosystem services as a valuable tool for dialogue and agreements about the equitable and sustainable sharing of 'common' catchment resources. They also provide a framework for determining market mechanisms that may include agreeing charges for the use of some ecosystem services (abstraction of fresh water, land use, etc.) and remuneration for active management to contribute to the production of services (such as flood regulation, habitat provision, land stewardship for water yield, carbon sequestration and wildlife conservation, and management of culturally valued landscapes).

To value or not to value?

As noted in Chapter 5, a utilitarian view of nature does not sit comfortably with all ecologists, conservationists and religious interests, and the concept of monetary valuation is also often regarded with suspicion.[25] The predominant argument is that valuation effectively imposes the human construct of money, determined on the basis of human utility and/or preference, upon living systems not only preceding the creation of the economy but also vital for its continuance. For this reason, there remains some resistance to the whole concept of ecosystem services among some nature conservation interests. In some instances this is with good reason, as there are critically endangered species and habitats to which we morally and intellectually should ascribe, in effect, infinite value lest these essential and irreplaceable generic resources are expunged from an increasingly human-impacted world.

However, it is useful here to recall the fundamental purpose of sustainable development and supporting economic methods, which is not to subjugate ecological and different cultural value systems to the forces of the current market but rather to integrate them into more far-sighted decisions normalized on a common basis. Pragmatically, if we fail to ascribe economic values or some other weighting of likely significance to land, landscapes, species, habitats and their actual or potential values to society, they are implicitly considered as of zero worth in decision-making processes largely shaped by the still-dominant market model predicated upon the liquidation of natural and other resources for rapid financial return. For all the understandable and justifiable concerns of some conservation and spiritual interests, economics provides a normalizing mechanism for balancing different aspects, bringing them into political and corporate decision-making processes rather than relying on altruism to weight a more diffuse and far from consensual 'inherent worth'. Thereby, ecosystems are placed in a far more central and important place in the decision-making process. A positive value is automatically better than no value when it comes to safeguarding the worth of supportive, productive and otherwise culturally prized landscapes and habitats in public policy and corporate decision-making. We can forgive the lack of absolute meaning in derived monetary values, given that they support rather than determine decision-making. Importantly, rather than merely force-fitting them into the market, valuation of the range of services and the perspectives of their diverse beneficiaries serves to develop the market progressively to further internalize the ecosystems that underpin both the economy and all other dimensions of human well-being.

Valuation in support of sustainable and equitable decision-making

Given the many assumptions required to link many services to surrogate markets, deduced values may be illustrative rather than precise. The integration of many such disparate valuation methods and dependent assumptions into a total economic value is subject to further substantial uncertainty. For this reason, absolute derived total economic values for ecosystem services tend to be highly unreliable, and are therefore inherently open to challenge from opposing interests. This is handled in practice by using marginal differences in value arising from changes in the use of land, management regimes or other development or policy decisions. Marginal valuation offers a substantially more reliable basis for ascribing net value because, by applying the same set of assumptions to both a 'baseline' status of ecosystem services and those affected, or likely to be affected, by an intervention, the same uncertainties are internalized in both sets of values, leading to a significant degree of 'cancelling out' in assessment of ensuing marginal changes. It is therefore generally accepted by environmental economists that, while 'absolute' economic values have no objective meaning owing to their sensitivity to the methods and implicit assumptions applied, 'marginal' benefits and costs deduced by comparison of pre- and post-intervention valuations are highly informative.

A practical example of how this approach has informed government thinking is provided by an excellent study on options for a large-scale dam and water transfer scheme in South Africa.[26] The South African government has a long history of intercepting the flow of water in rivers rising in Lesotho and/or the relatively damper south-eastern province of KwaZulu-Natal, and pumping or diverting it northwards to feed the demands of the nation's economic heartland in the otherwise arid province of Gauteng. However, with the transition to democratic government in 1994 and the subsequent reappraisal of social and environmental policies, significant questions were raised about the long-term sustainability and ethics of continuing uncritically with this heavy engineering approach. To support decisions around proposals for the damming and diversion of the waters of the major tributaries of the Thukela river system, the then Department of Water Affairs and Forestry (DWAF) commissioned a quantified economic assessment of the wider Thukela Water Project (TWP), which addressed different options for diverting the flows of the river system. The study used ecosystem services as a basis, focusing in particular on the most relevant of the MA suite of services but also adapting the approach better to reflect local livelihoods and circumstances. Ecosystem services targeted by the study included three categories related to fish (non-recreational fishing and recreational angling both for trout and in the estuary), reeds and sedges, waste assimilation, waste dilution, cultivated flood-plain agricultural land, livestock grazing, whitewater rafting and kayaking, canoeing and recreational swimming. Six potential ecological 'disservices' were also addressed, including both the treatment costs and loss of productivity associated with bilharzia, cholera and pathogens associated with diarrhoea.

Interviews with and surveys of households and tribal communities representative of livelihoods and demography in each of the major sub-catchments of the upper Thukela system provided basic economic inputs. Findings were extrapolated up into total economic values for each sub-catchment of the Thukela system. This was achieved by multiplying the household benefit or disbenefit by the number of households enjoying the ecosystem service within each sub-catchment. This process established baseline values for the suite of target ecosystem services across the river network. Then, the environmental flow regimes likely to result across the river network were modelled on the basis of the various options for damming and water diversion under the TWP. From the modelled changes in sediment and flow regimes, impacts upon target ecosystem services were calculated. Survey data were then rerun in valuation models on the basis of likely changes to ecosystem services under each damming and river diversion proposal.

Benefits and disbenefits were assessed both across the Thukela catchment as a whole and as they impacted different communities within constituent sub-catchments. These impacts included 'out of river' implications (changes to the volumes of water available for abstraction and direct economic use) and 'in-river' services (the implications of changes to the level of ecosystem services supplied by the volume of water that remains within the river). For both the 'status quo' and different environmental flow scenarios, data were incomplete and based on various assumptions. Total economic values deduced therefore had no absolute meaning, yet the marginal values deduced by comparison of 'status quo' and scenario values for each of the dam/diversion options, based on the same set of techniques and assumptions, served the purpose of deriving marginal valuations illustrating directions of change and their likely orders of magnitude for ecosystem services and their consequences (both beneficial or prejudicial) for different dependent communities in the main sub-catchments.

Marginal economic changes were compared to the status quo assessment by multiplying the scenario impact (i.e. 10 per cent reduction to fishery potential) by the assessment of current economic value, also taking account of the population of beneficiaries. This was determined for eight tributaries, with four scenarios (including the status quo as a base reference) using seventeen potential services (or disservices) that may change, with overall economic assessments as well as breakdown by service in each tributary: some 544 different conditions in all.

Significant changes in services were most often attributed to reduction of water volume and flow with associated habitat loss, possible salt-water intrusion in lower reaches of river, loss of reeds and sedges (or in some flow scenarios significant increases) with consequences for indigenous construction and crafts, impacts on the capacity of rivers to dilute and assimilate wastes, lowering of the water tables on flood plains, impacted flood-plain grazing, affected recreation and fishing opportunities and, often with highly significant impacts, increasing habitat available to vectors or direct transmission of debilitating waterborne diseases including bilharzia and cholera.

For many dam/diversion scenarios, a different balance of benefits was encountered in different sub-catchments across the river system. This demonstrates the hazards of too broad a generalization about impacts. Under some scenarios, disease implications arising from reduced dilution and the creation of flow regimes more amenable to parasite vectors were found to be particularly significant. A different balance between beneficiary communities and those suffering disbenefits was also revealed under each scenario. Commonly, there was a significant redistribution of benefits and disbenefits across different groups of people and their livelihoods, mediated by the ecosystem services within the river system upon which they depend. Some 'development' options were likely to result in significant disadvantages for some sectors of society, and occasionally for all people in sub-catchments.

Some of these often overlooked disbenefits can be significant when economic value is factored into total outcomes. For example, the spread of waterborne diseases such as bilharzia, cholera and diarrhoeal conditions may be accelerated not only by the creation of large bodies of standing water behind dams but also through reduced dilution effects and declining flows lower in the catchment which favour the spread of invertebrate disease vectors (a water snail in the case of bilharzia). The Thukela study provides evidence of potentially significant boosts in the incidence of bilharzia and its associated economic impacts throughout the river system under different dam development and water transfer scenarios. This mirrors the generally unanticipated yet widely observed and often devastating spread of malaria, West Nile fever and Japanese encephalitis across India consequent on dam construction, and the proliferation of waterborne diseases under hydrological modification elsewhere in the tropical world.[27]

The *Economic Impacts on Ecosystem Services* study of the upper Thukela river system summarizes its complex analysis simply for decision-makers, in both central and provincial government, through a 'traffic light' colour coding. This colour coding – pale blue (no problem); green (good); orange (caution); and red (bad) – spanned the full suite of ecosystem services across sub-catchments, making the likely consequences visible to all stakeholders. Therefore, no decision could then be made without commensurate exposure of its consequences for different communities. At the time of writing, no decision on development of the TWP is forthcoming in South Africa. Knowing what we do about this study into associated detriment to local ecosystems and human livelihoods, were a decision taken to divert water northwards to Gauteng then it would effectively and visibly perpetuate the hegemony of the dominant, predominantly white economy.

The Thukela study, as will be the case for all such 'real world' studies, necessarily simplifies the complexity of environmental processes. For example, it was founded upon changes in environmental flows and not broader ecosystem processes (sediment trapping, changing water-use habits, increased evapotranspiration, etc.) which may further influence total marginal costs. However, the study provides a robust and graphic demonstration of the interdependence of multiple ecosystem services and the way that they can impact sub-catchments and sectors of society in different ways under changed river management scenarios.

The Thukela study remains one of the most advanced in the world addressing the broader consequences of different development scenarios across a major river system. It suffices, for the purposes of this chapter, to highlight how valuation could inform more integrated decision-making when marginal values are deduced from comparison of current (baseline) state with a number of projected future states likely to result from various development options. It remains a pioneer of the kind of integrated and transparent assessment necessary to inform strategic decisions in the future, and to apply them more widely not only throughout Africa but across the globe. For those interested in pursuing this study, some simplified accounts contain more detail.[28]

Streamlining the ecosystems approach in landscape-use decisions

The current shortfall in the full application of the ecosystems approach is hardly surprising when one weighs its relatively recent consolidation and penetration into the policy arena, particularly since publication of the Millennium Ecosystem Assessment in 2004/05, against centuries of habit and inherited assumption based on reductive or holistic views of the world. Also, the resources required to undertake a fully detailed analysis of all services are significant, and may be prohibitive in terms of cost and time lag.

An extremely helpful approach to streamlining a systems-based view of the implications of development decisions was published by the UK's Department for Environment, Food and Rural Affairs (Defra) in its 2007 document 'An introductory guide to valuing ecosystem services'.[29] The Defra guide's approach is to undertake a preliminary pre-screening of the likely significance of impacts across the full suite of Millennium Ecosystem Assessment ecosystem services, first assessing the 'baseline' condition and then evaluating the marginal impact of different development options. This rapid screening is initially performed by expert opinion, ideally reflecting the consensus of a range of stakeholders, ascribing a likely significance of impact for each ecosystem services using the semi-quantitative scoring system in Table 7.2. This firstly ensures that relative significance is considered in relation to each development option to gain a mental picture of the balance of impacts across all ecosystem services and their respective beneficiaries compared to baseline (no development) condition. Secondly, it highlights the most significant or uncertain impacts upon ecosystem services, for which limited resources may be targeted for more detailed study. This two-phase approach retains the overall systems view and acknowledges real-world resource constraints, but prevents the latter compromising the former.

TABLE 7.2 Semi-quantitative scoring of the likely effect of development options for each MA ecosystem service

Score	Assessment of effect
++	Potential significant positive effect
+	Potential positive effect
0	Negligible effect
–	Potential negative effect
– –	Potential significant negative effect
?	Gaps in evidence/contention

Source: Defra (2007b[30])

Simple though this Defra weighting system may appear, it can nevertheless inform substantial conclusions. The weighting system, and not monetization, has formed the basis for a range of practical analyses. This includes, as three examples, its use to support decisions about the best options for managing an end-of-life coastal flood defence bank in the south of England (used as a good practice example in the Defra valuation guide),[31] the most sustainable development choices for sites in the east of England,[32] and the generation of politically influential conclusions about the likely implications of a proposed large dam scheme on the India/Nepal border.[33]

Thus far, this ecosystem-service-based appraisal may seem somewhat academic. However, insistence on a whole-system approach is far from pedantry, as real ecological, social and economic consequences can flow from a failure to consider the whole connected socio-ecological system. Without this whole-systems view, sustainability is the net victim as we revert to the traditional, established narrow 'exploitation economics' approach of favouring only selected services and beneficiaries over all others, which are entirely overlooked or considered unimportant. The following section of this chapter considers published case studies reporting upon ecosystem-service-based analyses of practical environmental management across the fields of flood risk management, protection of fresh water and river rehabilitation.

Evolving markets for sustainability

Transition from a current exploitation-based to an ecosystem-based market implies radical reform. While some observers may consider this an implausibly large leap, the transition, as outlined in the preceding examples, is already manifest in the emergence of markets for carbon sequestration as a result of the now increasingly evident consequences of overusing the 'common' of the atmosphere to dissipate waste gases. It is therefore entirely reasonable to expect the same scale of market innovation in the management of flood risk, water yield, fishery recruitment and other dimensions of the environment now that our understanding of human and economic interdependency with ecosystem services has gained so much momentum and political currency. The value of land for services other than limited food, fibre and amenity uses must necessarily evolve into more wide-ranging markets.

One such study exploring market development around ecosystem services has been undertaken on the Mvoti catchment of KwaZulu-Natal, South Africa.[34] The purpose of this study was to support a stakeholder engagement process with the ultimate objective of institutional reform to enable people in the catchment to share water more equitably and sustainably. From this work, it became clear that the innovation of social and institutional processes and their supporting economic arrangements to achieve these ends is at least as important as technological innovation. The key economic issue is to connect people and their livelihood aspirations both with each other and within a common understanding of how shared ecosystems integrate the needs of all stakeholders. This then opens the way to participatory governance and markets to protect the integrity of the whole socio-ecological system. The economics of a systemic world then revolve around protection and wise investment of this critical natural capital in as diligent a manner as financial capital is managed today,[35] acknowledging that the consequences of collapse of supportive ecosystems are overwhelmingly more grave than a mere economic 'credit crunch'.[36] Economics may then be reformed to become not an instrument to liquidate nature for rapid profit, as it has so often been used in our industrial past, but as a tool to help recognize, safeguard and share equitably the benefits of ecosystems. Environmental management costs are then construed less as an inconvenient inhibition on economic progress through taxation to support 'nice to have' ecological assets, but as investment in the systems underwriting future security and opportunity.

This is achieved when markets evolve progressively to encompass more of the ecosystem resources that underpin them. Quite how this natural capital-based market will eventually unfold remains at present uncertain. Both emerging and mature 'paying for ecosystem services' (PES) markets across the world, linking producers to users of selected ecosystem services, will be extensively discussed in the next chapter and may provide models for the future evolution of novel markets naturally attuned to the sustainable use and sharing of ecosystems and their disparate services.

Valuing the natural world; valuing all who share it

Whether in economic terms or simply in recognition of the broad range of dependencies that humanity and all of its activities have upon land, landscapes and broader ecosystems, valuation of ecosystem services is fundamental both to their sustainable stewardship and as a basis for sharing. Recognizing our complete reliance on the many functions of the natural world, whether in their wild state or concentrated through agriculture, water supply infrastructure or other management practices, we are also forced to confront the implications of their loss, degradation or annexation by competing interests.

An implicit morality stems from the biophysical reality of our sharing of common pools of ecological resources, broadly consistent with the 'golden rule' of not doing to others that which we would not have done to ourselves. Ecosystem services then provide us with mechanisms not merely to rethink what is implied by sustainability and how this ultimately serves our own best interests, but also requires of us increasingly equitable values and behaviours such that we learn how to share indefinitely that which makes life possible, profitable, enjoyable and fulfilling.

8 | Living landscapes

Relatively recent expansions of human population, technological capabilities and lifestyle demands have progressively overridden natural carrying capacities such that the degradation of ecosystems and their supportive abilities threatens to limit further human advancement.[1] Humanity has always depended upon the services of the landscapes it inhabits; this will always be so. The pressing challenge of sustainable development, then, is to apply our insights and understandings to plan for wiser uses of natural resources such that they can continue to support our multiple needs indefinitely. I refer to the achievement of this balance between economic and other livelihood uses by humans and the renewable capacities of ecosystems, such that all can continue in perpetuity, as attaining 'living landscapes'.

Given the complexity, embedded habits and vested interests of the modern world, active and far-sighted interventions will be essential to achieve this balance, rather than hoping that sustainability will just happen as a consequence of new awareness. In this regard, the unification of ecosystem functions with human benefits and economic values posits ecosystem services as one of the most potent frameworks for integrating different needs and value systems in the re-engineering of living landscapes.

Tools for systemic management of land

An abundance of studies based on ecosystem services have been conducted to address various environmental problems.[2] However, many studies today still commence with preselection of ecosystem services deemed to be either 'important' or which are relatively easy to quantify and/or monetize, and therefore by definition which are already internalized by today's market. As discussed at some length in previous chapters, this preselection of services overlooks the systemic intent of the ecosystem service framework by disregarding other connected services, their multiple beneficiaries and the interdependencies between them. When the linkages between services are lost, so too is the essence of the system, and we will almost inevitably achieve nothing more than maximizing the production of a slightly broader set of services already with market value (such as food and fibre production) while continuing to overlook impacts on other services currently external to the economy. At best, we will tend to fall back on the suboptimal holistic outcome of wilfully or, more commonly, inadvertently achieving 'trade-offs' between services rather than identifying 'win-win-win' innovations or alternatives based on a whole-ecosystem approach.

Converting powerful frameworks into pragmatic and consensual tools to inform day-to-day decisions remains a priority if we are to progress from good intentions to living with landscapes in ways that are equitable and sustainable, securing the multiple benefits that they confer upon all in society.

Reconnections

When we dig, drain, mine, harvest, inhabit or otherwise exploit landscapes, we inevitably influence the structure, functions and associated ecosystem services that they provide. All species do this. Most are limited in these activities by feedback mechanisms, such as populations of earthworms which are controlled by rising populations of predators if they proliferate beyond the balance point. Other species, such as corals and climax vegetation such as forest trees, can themselves create structures supporting associated ecosystems which have evolved to exist in balance.

Humanity differs to the extent that our application of knowledge through technology has enabled us to break free of many former natural constraints, such as food limitation, disease epidemics and climatic barriers. This innovation has allowed us to prosper geographically, numerically, technologically and in many other ways. However, the nature and scale of this proliferation are not without longer-term consequence. The relearning that is required at this critical time in our evolution is that all things are connected, and that the impacts of all our actions extend far beyond local scales of time and space and also beyond immediate target beneficiaries and outcomes. It is clear from an ecosystems perspective that these interventions must necessarily affect entire ecosystems, their services and their many beneficiaries at many scales from the local to the global. Impacts of one kind or another are inevitable, so the key issue is how the wider ecological and societal ramifications are considered in the decision-making process.

In this chapter, we will look at some of the reconnections that are being made in the fields of land management for flood risk, production of fresh water and also the restoration of river systems. These instances illustrate the kind of tools that we may need to innovate across other environmental disciplines if we are to attain a more sustainable relationship with the ecosystems essential to our future well-being.

Reconnections through flood risk management

Chapter 4 explored the progressive transition across much of the developed world from a narrow disciplinary and geographic focus on 'flood defence' concepts and practices towards the new systemic paradigm of 'flood risk management'. Flood risk management considers the interaction of water with whole catchment landscapes and human uses. It does not seek automatically to rid the land of water, but recognizes the value of water naturally detained in flood plains and interacting with groundwater as a contribution to the smoothing of catchment hydrology and the minimization of flood risk to settlements and land downstream. This inherently acknowledges ecosystem services, particularly those arising from hydrological processes, provided by critical habitat within catchment ecosystems. Protection or restoration of these services by sympathetic land uses begins the reconnection of societal benefits and economic values with ecosystem functions. By focusing on the source of flood risk rather than 'defending' against local downstream impacts, outcomes are likely to be more economically as well as ecologically efficient. However, acceptance of this new approach, particularly the 'surrender' of land that is now allowed to flood again periodically as part of restored natural regimes, may jar with the expectations of land-owners and local people regarding land 'defended' throughout their lifetimes, albeit on an increasingly uneconomic basis. Two English examples illustrate how an ecosystems perspective can result in more beneficial, equitable and socially acceptable outcomes from flood risk management decisions.

The first example is an ecosystem service-based appraisal commissioned by the government department, Defra.[3] This study explored options to manage 20 kilometres of flood banks in poor and deteriorating condition at the western end of Poole Harbour around the town of Wareham (Dorset, UK). The risk of overtopping or breaching of these defences was aggravated by rising sea levels, with the loss of important intertidal habitat exacerbated by 'coastal squeeze' elsewhere in the estuary. Ecosystem services were used to assess likely outcomes from a range of management options addressed, including provisioning services (fisheries), regulating services (nutrient cycling, carbon storage) and cultural services (recreation, archaeology). The responses considered included 'do nothing'; 'do minimum' (continue existing maintenance, which would merely delay defence failure); 'improve' (maintain appropriate standards of defence in the face of sea-level rise); or three variants of 'managed realignment' (a preliminary vision, the staged removal of tidal banks, and full removal of tidal banks with some secondary defences provided for a small number of selected key habitats that could not be readily re-created elsewhere). The first pass of policy appraisal did not seek to apply economic values to any environmental impacts, consistent with what would become Defra's risk-based scoring approach as outlined in Table 7.2 in Chapter 7. This first pass concluded that the managed realignment scenario entailing removal of flood banks appeared the most promising both as a flood defence solution and as a means to create habitat to compensate for that lost elsewhere in the ecologically important Poole Harbour.

Further work was then carried out to determine the costs and benefits associated with changes to ecosystem services for the options, based on detailed physical, ecological and broader environmental analyses. This revealed that maintaining defences was very unlikely to be desirable as it yielded significantly lower ecosystem service values and higher engineering and maintenance costs than all other options. The 'do nothing' option displayed some high ecosystem service benefits although these would be substantially offset by flood damage costs, various other risks and the likely loss of recreational values. 'Do minimum' was shown to have slightly lower ecosystem service benefits, but higher maintenance costs. However, the most strongly beneficial option was found to be managed realignment, with the removal of flood banks topping the list.

The Wareham study was significant in that it placed ecosystem services – not all of which were monetized or free from data gaps and uncertainties but all of which were at least risk-assessed – centrally in decision support. This directly informed the management outcome (managed realignment with protection of some valued habitats) on the basis of what was likely to deliver the greatest and most enduring public value. The Wareham study also directly informed the shift in government policy, as exemplified by its inclusion as a case study in the Defra ecosystem services valuation guide[4] and its direct contribution to decisions and public explanation about a substantial flood risk investment decision. The case study also demonstrates that 'perfect' ecosystem service valuation is not essential for many appraisal purposes, which may simply require comparison of the relative magnitude of changes in the provision of ecosystem services across different options. Provided these are based on a clear audit trail of assumptions and, ideally, involve the participation of key stakeholders, semi-quantitative assessment may be sufficient for defensible decision-making purposes.

Reconnecting land and water to address flood risk

The second flood risk example of transition to a more systemic approach is illustrated by an ecosystem services study of the Alkborough Flats managed realignment scheme.[5] The Alkborough Flats scheme, on the south bank of the inner Humber estuary at the confluence of the River Ouse and the River Trent, is one of the largest managed realignment sites and flood storage schemes in Europe on 440 hectares of flood plain lying below the village of Alkborough in North Lincolnshire. It represents £10.2 million of multi-objective investment, including flood risk management and biodiversity benefits as well as social and economic benefits to the local community, while maintaining the navigability of the Humber estuary and the viability of local farms affected by the change of land use. Controlled breaching of former uneconomic flood banks allowing tidal inundation of Alkborough Flats inland as far as much smaller newly created bunds provides substantial flood storage area, calculated to be sufficient to reduce high tide levels over a large part of the upper estuary by 150 millimetres (with a pessimistic estimate of 100 millimetres). At a projected annual sea level rise of 4 millimetres per year until 2025 and then 8.5 millimetres per year until 2055, the Alkborough Flats scheme therefore modifies the regime to account for perhaps twenty-five years of climate change impact. It also defers improvements to other flood defences in the tidal rivers upstream of the site which would otherwise be needed to counter the effects of sea level rise, saving many millions of pounds of public investment. Managed realignment at Alkborough Flats also creates important intertidal habitat, supporting the internationally important bird populations and wider biodiversity of the Humber estuary and creating long-term recreational opportunities.

It was initially expected that there would be a net loss of provisioning services (surrender of arable production of food and fibre) 'traded off' to enhance regulatory (flood management was a key design outcome), cultural (birdwatching and other amenity activities) and supporting (provision of habitat was the other major driver of this scheme) services. However, valuation of outcomes from managed realignment in the ecosystem services assessment, subsequently integrated into a total benefit value and compared to scheme cost, revealed a number of important lessons. Perhaps unsurprisingly, annualized enhancement of the key services implicit in scheme design – 'provision of habitat' and 'natural hazard regulation' – were found to be significant. However, additional significant benefits accrued from the cultural service of 'tourism and recreation' and the regulatory service of 'climate regulation'.

A significant and surprising finding of the study was that, contrary to initial expectations, the net outcome for provisioning services was effectively neutral and possibly significantly positive. The net loss of £28,075 for 'food' production (loss of arable food crops partly offset by sale of meat) was largely compensated by the 'rare breeds' grazing regime that replaced it, yielding a £26,820 gain for fibre production (wool production minus loss of straw production) with a further estimated £3,000 gain from genetic resources (sale of rare breeds stock). The study recognized as a key knowledge gap that it was not possible to quantify and therefore monetize the recruitment of fish species with substantial commercial and recreational value which, if reliable values could have been deduced, were likely to contribute to a substantial net uplift in provisioning services. When all quantifiable ecosystem services were compared to the costs of the scheme, a benefit-to-cost ratio of 3.22 was calculated over scheme life. This recognition of the wider outcomes of the management realignment naturally resulted in a higher benefit-to-cost calculation relative to the more traditionally and narrowly focused benefit assessment exercise undertaken initially to justify the scheme (which had produced a benefit-to-cost ratio of 2.72).

Aside from the economic outcomes, a wide range of additional lessons were also learnt from the Wareham and Alkborough Flats ecosystem services case studies. These included the value of the ecosystems approach in addressing land and landscape uses to resolve complex socio-ecological challenges such as those posed by flood risk management. The neutral or positive impact for provisioning services provides an unanticipated and striking example of how, when the full suite of services is assessed within a contiguous system, innovations can be identified – in this case in the alternative stewardship of land – to optimize the balance of public benefits flowing from land and landscapes. An automatic assumption that there will be a 'trade-off' blocks innovation for possible 'win-win' outcomes.

The implications of this finding for changes in agricultural subsidy and policy are significant and transferable to other nations and bioregions. They need to be championed if they are to result in revision of agricultural policies and subsidies, recognizing the potentially significant contribution of land use to multiple aspects of public value contributing to overall sustainability. This can justify subsidies for land use on the basis of their intended beneficial outcomes for all in society, which is a policy intent of the 2003 reform of the EU Common Agricultural Policy as reviewed in Chapter 2. The language of ecosystem services, founded as it is upon the benefits accruing to different sectors of society, provides a helpful basis for communicating the intentions of ambitious management schemes and, where there may be clear winners and losers from changing management, enables negotiation about compensation or other forms of sharing of associated benefits and burdens. It also demonstrably provides a mechanism through which to consider innovations that may overcome negative impacts as compared to traditional 'trade-offs'.

Reconnecting land use with production of fresh water

We have already explored the adverse and unintended consequences of modern industrial agricultural systems and the inherited market economy that drives it, founded as they are on the reductive paradigm of maximizing food and commodity yield per unit of land area. The largely unintended degradation of landscape quality, the pollution, eutrophication and siltation of connected surface water, wetland and groundwater bodies, deterioration of habitat and species, and connected human uses such as the quality of abstracted water and the vitality of fisheries, represent real but largely externalized costs. The UK government's 2002 'Curry Report' on *The Future of Food and Farming*[6] began to challenge inherited assumptions that land use should be directed solely at the production of food and/or other commodities, and that there should instead be broader emphasis on the wide range of additional publicly beneficial functions of land and landscapes.

The management of land for other services, significantly including stewardship of catchment landscapes for the 'provisioning service' of fresh water, can have substantial associated societal and market benefits. This is exemplified in four case studies of land co-management for food and water production from the USA, France and England.

The first of these case studies is the water supply of the city of New York, which is a dramatic, large-scale example of the management of land and landscapes for the benefit of water supply.[7] Provision of adequate, safe drinking water to any city is a vast and complex undertaking. New York City's Department of Environmental Protection delivers over 1.2 billion US gallons (4.5 billion litres) of water daily to 9 million people. This water supply developed in a piecemeal fashion as the city grew, with the mushrooming population throughout the nineteenth century necessitating development of a network of aqueducts and reservoirs bringing in water from sources of high natural quality from rivers and wetlands draining from a considerable distance to the north of the city. In 1905, New York City looked northwards again in its quest for more high-quality water, identifying the Catskills Mountains as a prime resource and building various dams and reservoirs throughout the Catskills up to

1928. Faced with ever-growing demand, the city had also turned in 1927 to sources in Delaware County. After some legal wrangles, construction of the Delaware component of the Catskills–Delaware system, known collectively as the Cat/Del system, was implemented in stages between 1937 and 1964. Today, New York City has the largest unfiltered surface water supply in the world, delivered by natural capture and purification processes throughout a catchment of 2,000 square miles (830,000 hectares). The economic value of this ecosystem service is substantial compared to the costs of abstracting, pumping, storing and purifying water from the contaminated and enriched Hudson and other lowland rivers close to New York City.

Traditionally, the city had relied on sparsely populated rural catchments and low-intensity land use across the Catskills to protect raw water quality. Yet, by the 1980s, industrial-scale agriculture was replacing traditional methods and residential development added to the threats in these environmentally vulnerable areas. At the time, less than 30 per cent of the watershed area ultimately supplying the city was in public ownership, increasing potential risks to the quality of raw water from the actions of intensive agriculture, industry, highways, residential areas, forestry and tourism activities. Simultaneously, public health standards were becoming more stringent, requiring that all public water supply systems provide filtration or else meet a higher level of water quality, operational and catchment management targets.

Faced with the potential need to implement expensive filtration plant at an estimated cost of between $4 and $6 billion at 1990 prices,[8] plus annual running costs of over $200 million, the city's planners began to think in broader terms about protecting the resource. New York City initiated a comprehensive 'watershed protection programme' with mandatory standards for raw water, service to customers and source management. This comprehensive programme of watershed protection compared favourably on a cost and effect assessment when contrasted with physical and chemical filtration, though the practicalities of making it work would be daunting. Top-down, punitive regulations to force such systems to work had generally ended in failure elsewhere. For this reason, the city opted for a mutually beneficial, urban–rural watershed protection partnership, providing simultaneous benefits to the residents of New York City and the communities in the Catskills and Delaware catchments which manage the water-yielding landscape delivering the provisioning service of 'fresh water' enjoyed by urban residents. Through a process of dialogue and consensus-building, farmers were educated about the environmental and economic risks associated with some farming methods while, in turn, the farmers educated city representatives about the economic pressures they faced and how previous anti-pollution measures were not workable. This urban–rural watershed protection partnership arrangement came into operation between the city and the farmers by the end of 1991, including development of Whole Farm Plans to integrate agricultural pollution control into individual farm businesses. These Whole Farm Plans were developed in partnership with a range of statutory natural resources, soil and water conservation bodies, as well as agri-environment and other agencies and the farmers themselves, representing a consortium of common interests in the minimization of environmental risks. Staff inputs and the capital costs of pollution control investments on each farm were paid for by the city as an incentive for farmers to join the programme. The process was completed in January 1997, when the constituent parties formalized all prior agreements within a comprehensive memorandum of agreement, to which the city committed funds of approximately $350 million in addition to the costs of various other initiatives in the watershed. Parallel partnership initiatives have since been implemented between the city and forestry interests, land acquisition programmes and ecologically based land management. The total cost of the watershed protection programme is approximately $1.3 billion, substantially less than that of the filtration plant, and will maintain the city's pristine water quality for the foreseeable future. Within five years, 93 per cent of farmers in the Catskills had joined the Whole Farm Plan programme, increasing to 95 per cent by 2008. This reduced agricultural pollution by 75 per cent and stabilized the economics of farming in the Catskills.

The New York example demonstrates an emerging practice of 'payment for ecosystem services' (PES), which has been developing around the world at least since the early 1990s. PES is a form of market for ecosystem services comprising '... a voluntary, conditional agreement between at least one "seller" and one "buyer" over a well defined environmental service – or a land use presumed to produce that service'.[9] PES schemes can create, and have formed the basis for, self-reinforcing markets to secure the ecosystem services of water production, nature conservation and carbon sequestration.[10] A 2010 review by the OECD suggests that there may already have been 300 operating PES schemes across the world.[11] Prerequisites for successful PES schemes include a robust understanding of the stewards of land and other resources 'producing' a service (or group of services), of the actions required for an 'additionality' of service production beyond legally required minima, 'buyers' willing to invest in that service, and strong governance to control the market.[12] The scale and efficacy of the New York City water supply scheme makes it a global exemplar of the PES approach. Water-related services from catchments commonly form strong candidates for PES schemes, but the services provided by forests (including water resources, flood control, amenity, carbon sequestration and biodiversity) are also well studied and amenable to market-based instruments.[13]

Freshening the spring

The second case study is of the Vittel spring in France,[14] which demonstrates the potential role of markets for catchment services. Water marketed under the Vittel brand is drawn from the *'Grande Source'* ('Great Spring') located in the town of Vittel at the foot of the Vosges Mountains in north-eastern France. This source, from which water is naturally lifted from a 6,000-hectare aquifer some eighty metres underground, has been ascribed beneficial properties since Gallo-Roman times. The spring passed into family ownership in the 1840s and a spa was developed in the town of Vittel, attracting visitors from all over Europe. Bottling and marketing the water developed from 1882, at which time the Vittel brand was created. One million bottles had been sold by 1898, and 3 million bottles of water were sold in 1903. By 2006, 1 billion bottles of Vittel mineral water were sold annually across seventy countries. The spring was progressively purchased by the multinational Nestlé group between 1969 and 1992. Maintenance of quality is vital to the water bottling business owing not merely to reputation risks but because, under French law, 'natural mineral water' is required by legislation to be bottled at source from a well-protected, specific underground supply of stable quality and with no further treatment other than the elimination of traces of potentially problematic metals. To be labelled 'Vittel', the water cannot contain more than 4.5 milligrammes of nitrates per litre and must not contain detectable pesticides. Failure to meet these conditions would mean that the right to use the 'natural mineral water' label would be lost, along with its substantial associated markets. (Treatment is authorized in some other countries, including the UK the and USA, which reduces business risk.)

In the early 1980s, the then family majority owners of the Vittel brand recognized that intensification of agriculture, particularly the progressive replacement of traditional hay-based cattle ranching systems with maize-based systems and an increase in stocking rates, posed a risk to nitrate and pesticides levels in the *Grande Source*. This led the family to consider five options for ensuring water quality over the next fifty years. These were: (1) 'doing nothing', which was considered too costly and risky; (2) relocating to a new catchment where risks were lower but which would mean losing the Vittel label and its premium value; or (3) purchasing all lands in the catchment, which was not feasible, as French legislation does not allow sale of agricultural land for non-agricultural purpose and Vittel would not anyhow have the capacity to manage all of this land. Furthermore, social protests were also likely if too much land were to be sold to non-farmers. The fourth option was to (4) require farmers to change their practices through legal action, though this lacked legal justification as the nitrate levels of concern for spring water lay within limits defined as acceptable. The final and, in practice, the only practical option was to (5) provide incentives to farmers to voluntarily change their practices.

This voluntary option raised the substantial challenge of finding ways to make Vittel's and farmers' interests coincide in ways that induced farmers to cooperate. Vittel's initial proposals in 1988 to farmers with land in the protection perimeter, though scientifically founded, was not accepted as farmers felt that it was not adapted to their production systems. This then initiated a ten-year process of dialogue to transform the conflict between farming and bottling interests into a successful partnership. In 1989, Vittel formed a four-year, multidisciplinary action research partnership programme with the French National Agronomic Institute (INRA). This research programme, known as 'Agriculture-Environnement-Vittel' (AGREV), had the three objectives of understanding nitrate pollution pathways, initiating changes in farming practices necessary to control this pollution, and identification of incentives necessary for farmers to change their practices. This research, which engaged both Vittel and the farmers, exposed the diversity of farm sizes, farmer age groups, farmed outputs and levels of indebtedness, all of which affected farming attitudes to changing practice. This helped identify the highest-risk groups of farmers, particularly those who might be persuaded to give up maize cultivation for animal feed in favour of extensive cattle ranching. These measures were radical compared to modern intensive systems, marking a return to more land- and labour-intensive practices in addition to significant investment. Farmer participation in action research under the AGREV programme helped development of a package of attractive incentives compatible with the goals of both groups of stakeholders over a period of ten years, simultaneously addressing land, labour and capital shortages. Farmers' interests were further served by Nestlé Waters being a major employer in the basin, which otherwise suffered high levels of unemployment, including providing jobs for family members in many of the farms. The process took a major step forward when, in 1992, Nestlé Waters (by then outright owner of Vittel) created Agrivair, an intermediary responsible for negotiating and implementing the programme, which built continuity between the design and the implementation of the project and was critical to facilitating the communication of the results of scientific research to the farmers. The resulting package of incentives developed and agreed in collaboration with farmers included: (1) long-term security through eighteen- or thirty-year contracts; (2) abolition of debt linked to land acquisition with land acquired by Vittel left in usufruct for up to thirty years; (3) subsidy of, on average, about €200 per hectare/year over five years to ensure a guaranteed income during the transition period; (4) up to €150,000 per farm to cover the cost of all new farm equipment and building modernization; (5) free labour to apply compost in farmers' fields, addressing the 'labour bottleneck'; and (6) free technical assistance. Contracts were agreed on an individual basis with participating farmers, with Agrivair also purchasing some land available in the region, particularly as farmers in the non-target small-farms groups chose to retire. By 2004, all twenty-six target farms in the area had adopted the new grazing regime, covering 92 per cent of the sub-basin.

The Vittel case study included a clear market between one buyer (Nestlé Waters) and twenty-six voluntary 'sellers' (participating farmers in farms targeted as the most likely to make the greatest difference to water quality), implemented under the leadership of an intermediary institution. Establishing this exemplary PES scheme at Vittel was a complex and necessarily slow undertaking, taking account not merely of scientific information but also of social, economic, political, institutional and power relationships, with maintenance of farmer income crucial to the transition. Trust-building through the long engagement process leading to co-creation of mutually beneficial solutions was seen as essential to success, which was achieved in the absence of complete knowledge of costs and benefits.

Nestlé Waters has replicated the Vittel source protection PES approach at the Perrier source in the South of France and the Contrex source in the *département* of Vosges, both of which it also now owns, achieving broadly similar outcomes. However, successful though this PES initiative has been, it is not included as a discrete case study as parallel lessons emerge.

Farming for nature and water

The third case study of land management for water production is one from the UK, addressing a farm-scale transition in stewardship at

High Hullockhowe Farm, in the uplands of Cumbria. Though a small farm, this provides a microcosm of how broad thinking can maximize benefits across a range of ecosystem services that significantly include the production of fresh water benefiting people remote from the farm.[15]

Owing to the foresight of those establishing the forerunners of to-day's water service companies in the UK, considerable swathes of generally poorly productive uplands were taken into municipal (now passed into private) ownership for the purpose of protecting sources of water valuable for public and industrial supply. High Hullockhowe Farm sits amid a wider tract of land in the ownership of the multi-utility operator United Utilities, which is both the farm's tenant as well as the water service company providing potable water to much of the north-west of England. Although unproductive, small-scale upland hill farms have historically benefited from subsidies to support and intensify activities, which have in turn had adverse consequences for the functioning of farming landscapes, including impacts on biodiversity and water sources at the top of the watershed.

Throughout the 1980s, upland farming practices at High Hullock-howe Farm, encouraged by regionally applicable subsidy systems, had been found to contribute significantly to livestock-related water pollution problems serious enough to force the closure of a water abstraction point on the Toddle Beck draining from the farm. The economic consequences of this closure were significant for United Utilities, vastly exceeding the small rental paid by the tenant farmer. Between 1994 and 1999, renegotiation of management arrangements, the tenancy agreement and direct support from the landlord for certain infrastructure improvements resulted in substantially different farm stewardship arrangements at High Hullockhowe Farm. These included development of a whole-farm plan, as well as the farm joining the then current Environmentally Sensitive Area agri-environment subsidy scheme, ceasing to keep cattle in winter housing, the fencing off and restoration of rush-dominated spring habitat, which also represented a hazard during lambing, and replanting of some stands of trees as windbreaks and also to provide additional habitat. New patterns of land management were also implemented both on the farm and in adjacent areas, with additional guidance from the Royal Society for the Protection of Birds (RSPB), including the re-establishment of pockets of arable land within a predominantly pastoral landscape to provide nesting habitat and food availability during the winter for a range of bird species including lapwing (*Vanellus vanellus*).

The primary interest of United Utilities was protection of water resources, effectively 'farming' the land for the benefit of the abstraction point lower in the catchment, but this of course depended on the restoration of the integrity and functioning of upland habitat critical to the service of fresh water production, which provided the coincidental opportunity for further nature conservation gains.

The costs of changing management regimes at High Hullockhowe Farm were met from various sources including the landlord (United Utilities), agri-environmental grants and savings in farm management. Net benefits were significant, including recovery of an annual (2004) value of £19,500 for the Toddle Beck abstraction point (a conservative figure excluding the marginally higher value of this water source during drier times of the year), which then exceeded the rental value of High Hullockhowe Farm by a factor in excess of six. Further 'intangible' benefits included improved populations and breeding success of target bird species, improved plant diversity, security of farm income, the attractiveness of the landscape with associated tourism benefits, and reduced risks of disruption to water supplies. Though the regime benefited the tenant farmer, United Utilities was both the major beneficiary and main promoter of management for fresh water production on its owned land, considerably facilitating implementation of this pioneering PES scheme.

Extension to landscape scale

The High Hullockhowe farm example was influential in the shaping of the fourth case study, SCaMP (the Sustainable Catchment Management Programme), covering a far more extensive area of upland in the north-west of England. Applying these lessons, United Utilities began to take a broader view of the management of its landholdings in the north-west of England, from which SCaMP was developed in collaboration with the RSPB from around 2002. SCaMP represented a more strategic and ultimately more cost-beneficial approach to investment in 57,000 hectares (140,850 acres) of its upland holdings.[16] All of this land is a source for the water entering reservoirs and rivers, which constitutes the primary resource for the utility business, but much of the land also supports nationally significant habitats and species, with around 30 per cent of the total area designated as Sites of Special Scientific Interest (SSSIs). United Utilities and its predecessor companies already had an active catchment management programme, recognizing the role of biodiversity and high-quality landscapes in producing high-quality water. However, SCaMP sought to develop this connection substantially.

The charges that English and Welsh water utility companies are permitted to levy for water services, and the proportion of investment permitted for 'environmental schemes', are controlled by regulation under the Asset Management Planning (AMP) framework, which operates on a five-yearly cycle. Negotiation with government and regulators for the first round of SCaMP took place under AMP agreements to run between 2005 and 2010, with initial proposals for investment in the restoration of habitat in four upland zones controlled by United Utilities critical for production of high-quality water and provision of habitat for rare species including the hen harrier (*Circus cyaneus*).

AMP funding was finally approved for only two of these upland areas, Bowland and the Peak District. Each of these represented a significant area that was also subject to substantial ecological degradation, with implications for water quality, regional tourism and farm incomes. SCaMP progressed in these areas as an integrated approach to catchment management, simultaneously addressing government targets for SSSIs and wider enhancement of biodiversity, a sustainable future for the company's agricultural tenants, and the protection and improvement of water quality. AMP funding was set aside for restoring seriously degraded peatlands, planting woodlands, fencing off vulnerable watercourses, moving lambing off the moorland and developing Integrated Farm Management Plans for more environmentally sympathetic land management. All of this entailed close working between United Utilities staff, tenant farmers and the RSPB, in many cases also involving renegotiation of tenancy agreements and support with applications for agri-environment grants. Funding was also available from United Utilities to invest in one-off infrastructure improvements, targeted planting of broad-leafed trees, and selective monitoring of biodiversity and water quality to determine the success of the various schemes under the first phase of SCaMP.

A significant element of the overall programme was the restoration of blanket bog on the moorland top, including various projects to block moorland 'grips' (drainage channels cut into the peat) that were draining the moorland with loss of characteristic habitat and oxidation of humic matter, which creates coloration problems in water abstracted downstream. By taking out this coloration at source, through revegetation of severely damaged peat and the rewetting of the moorland and moss regeneration, blanket bog habitat is being restored with all of its attendant ecological benefits while simultaneously delivering more reliable flows of higher-quality water.

Regeneration of upland ecosystems has become a long-term solution to the supply of high-quality fresh water and the recovery of wildlife, with results expected to accrue over the long term rather than immediately. For this reason, the full realization of benefits for water quality, hydrology, ecology, habitat, fishery recruitment, landscape and associated tourism and other benefits has yet to be determined, although early indications from hydrological, water quality and biological monitoring are encouraging.[17] This suggests that SCaMP represents a value for water service customer (i.e. public) reinvestment of approximately £10 million spread over five years, supplemented by relevant agri-environment subsidies. For this reason, SCaMP has been widely cited as an exemplar of a 'win-win-win' scenario for ecosystem-based investment for the benefit of biodiversity, society and the economy.

Reconnecting land management with water, wildlife and people

All four of these water management case studies – New York City, Vittel (and Perrier and Contrex), High Hullockhowe and SCaMP – have succeeded for a range of reasons. For each, the partnership-based approach was key to success, entailing involvement from and benefits to all key players. Each also had vested interests both in terms of 'providers', acting through restoration of ecosystems delivering the service of fresh water production, and 'buyers' through market mechanisms and appropriate institutions. The advantages to both parties from exceeding statutory requirements were also clear. For each, there was also a clear brokerage between the providers and sellers, and the means by which the product in this market (fresh water) was enhanced were also well understood. They thereby satisfy the key criteria identified for successful PES schemes.

Each of the examples also coincidentally yields a wide range of other ecosystem-mediated benefits, which happen fortuitously outside of the principal project aim. They thereby exemplify the substantial public value that can accrue from broadening the concept of farming from production of commodities alone into a model that recognizes the value of broader ecosystem services. In these instances, this includes water production and biodiversity, but other PES schemes around the world include management of carbon, erosion and air quality, the aesthetics of the landscape, erosion control and other ecosystem services.[18]

These ecosystem-centred solutions are demonstrably more sustainable than technological 'fixes' to address downstream problems, which tend to be based on 'heavy engineering' solutions with associated inputs of chemicals and energy and their associated climate-active and other polluting outputs which can contribute to broader environmental problems. The solutions adopted in the case studies are also inherently more equitable, rewarding stewards of the land, controlling rising raw water treatment costs for downstream water service customers, and recognizing the interests of all ecosystem service beneficiaries.

Reconnections through river restoration

River restoration has emerged as a means to redress the impacts of systemic catchment degradation through urbanization, industrialized agriculture, unsympathetic land drainage and 'flood defence', and a range of other pressures. River restoration schemes have often focused on habitat structure, sometimes with specific nature conservation, amenity (including recreational angling in particular) or urban regeneration objectives. Many historic river restoration schemes have had the objective of restoring rivers to some notional 'original state', although as we have seen this very concept is largely arbitrary.

Advances in the ecosystem approach have subsequently broadened recognition of the multiple functions performed by catchment landscapes, river systems and the many societal benefits that they provide.[19] Today, river restoration tends to address a range of services that may include provision of habitat, amenity and recreation, flood regulation and water quality. Obviously, this strategic intent is commonly constrained in practice by the length of river available for restoration, the extent of flood plain available for renaturalization, competing demands on use of the river and also the need to safeguard some features, such as flood management for buildings and infrastructure that may not be ideally placed. However, the intent is always to work with and ideally restore natural processes, including flow and habitat diversity, ecological successions and natural flood-water detention. Often, this is brought about through removal of historic reinforced riverbanks, hard flood defence structures or other engineering unsympathetic to natural processes, and frequently the rehabilitation of habitat features such as meanders, flood plains allowed to be inundated during periods of high flow, and riparian wetland systems, all of which deliver multiple societal benefits.

A number of prominent voluntary organizations have championed river restoration in the UK, latterly embracing ecosystem services as a framework to optimize and communicate wider benefits to society. These organizations include the River Restoration Centre,[20] a network of active and effective river trusts, and the Salmon and Trout Association.[21] The river trusts movement in particular was quick to realize that protection or enhancement of catchment landscapes had many societal benefits beyond what were often perceived to be narrow benefits to fisheries or nature conservation. Indeed, river trusts such as the Wye and Usk Foundation[22] (and its forerunner bodies) and the Westcountry Rivers

Trust[23] (WRT, founded in 1995) have accessed significant amounts of EU regional development funding in recognition that better stewardship of farm businesses has regional economic impacts not only through farm savings and diversification but by stimulating ecotourism and recreational angling on the back of enhanced river ecosystems. The river trusts have proved the ecological, economic and social value of taking an ecosystem service-based approach.[24] The Salmon and Trout Association was formed in 1903, but reconstituted itself under a charitable remit in 2008 with the mission statement 'Game anglers for fish, people, the environment' in recognition that the presence of healthy populations of salmonid fishes represent wider dimensions of public value than merely angling exploitation.

One of the substantial EU-funded initiatives undertaken by the WRT was Tamar 2000, which took a catchment-wide approach to enhancing the river ecosystem as a central element of stabilization of the regional economy. The River Tamar system is a predominantly rural catchment of approximately eighty kilometres in length in the south-west of England, draining to sea at the city of Plymouth, which is the only substantial conurbation in the catchment. Tamar 2000 entailed a series of farm visits by WRT staff to work with farm businesses to identify diversification options but also costs savings that were simultaneously beneficial to averting water pollution or habitat loss. Measures, including fencing off vulnerable wetland systems, separation of 'clean' rainwater from 'dirty' farmyard washings requiring treatment, reductions in pesticide and fertilizer application, and development of fisheries, were of simultaneous benefit to farm incomes and river quality, which in turn was a primary resource for the region's ecotourism industry.

Upon completion in 2000, reports were published on the wetlands of the Tamar system[25] and also post-project economic evaluation of the scheme.[26] However, a new study was published in 2009 to examine the wider outcomes of Tamar 2000 based on the MA ecosystem services framework, which had been published in the interim. This new ecosystem services study[27] found that farm improvements and diversification had yielded many uplifts to ecosystem services across all four categories: provisioning, regulatory, cultural and supporting. The economic benefits determined for these ecosystem services were substantial and, when compared to the costs of the scheme, contributed to a benefit-to-cost ratio of 109. (A benefit-to-cost ratio of just 6.4 had been determined in the more traditional and therefore narrowly focused post-project economic appraisal.[28]) This is, by any standards, a highly significant outcome underwriting the substantial and multiple societal benefits that can stem from ecosystem-based solutions to many environmental management problems. Among many learning points emerging from the Tamar 2000 ecosystem services case study was endorsement that seeking solutions through restoration of ecosystem functions was likely to yield benefits across all ecosystem service categories and to many beneficiaries. Even prior to publication of the Millennium Ecosystem Assessment with its new classification of ecosystem services, and the appraisal of scheme outcomes based upon it, the Tamar 2000 project had already demonstrated the value of founding management and restoration on ecosystem services, with the Convention on Biological Diversity citing Tamar 2000 specifically as the UK's example of the successful implementation of the ecosystem approach.[29]

Generic conclusions about the multiple beneficial outcomes of Tamar 2000 have been confirmed by a range of comparable English river-related ecosystem services case studies. These include the Alkborough Flats scheme, discussed previously, and assessments of outcomes from a sea trout restoration on the River Glaven catchment in north Norfolk[30] and buffer zone installation on a formerly severely trampled bank of the upper Bristol Avon in Wiltshire,[31] from which the vast majority of benefits (99 per cent and 90 per cent respectively) accrue to beneficiaries beyond the fishery interests instigating the schemes. In all cases, payback to society is substantial, potentially forming the basis for new approaches to tackling environmental problems in ways that coincidentally deliver benefits to more people simultaneously via a range of connected ecosystem services. Achieving this, of course, will require a more enlightened model of regulation and funding, which today remains often constrained by its heritage in narrowly channelled 'fixes' to address specific discipline- and beneficiary-focused problems at more parochial scales of space and time. However, the scientific and economic evidence is compelling and is augmented by a growing body of case studies from around the world, collectively demonstrating the efficacy and public value of the ecosystem approach.

Creating markets for ecosystem services in living landscapes

The PES model is gaining traction across the world as a means to secure socially beneficial ecosystem services in an economically viable way within living landscapes. South Africa's innovative democratic-era water laws in particular have prompted a range of pioneering approaches to adaptive governance, accounting for the implications for a range of ecosystem services in development decisions. Practical initiatives addressed elsewhere in this book include the monetization of ecosystem services in the Thukela *Economic Impacts on Ecosystem Services* study[32] (outlined in Chapter 7) and other practical planning guides referred to in the final chapter, 'The people's land'. The Thukela study also demonstrates how a comprehensive ecosystem services analysis can provide a basis for negotiation about relative rights and interests, bringing equity issues to the fore in policy dialogue and also creating a basis for innovation of solutions to maximize public benefit. An additional study in 2007 by the Maloti Drakensberg Transfrontier Project[33] sought explicitly to establish market linkages between the restoration and management of upper catchment areas of the Thukela and the Umzimvubu river systems for the purposes of stabilizing river base flows and quality, yielding economic benefits to communities lower down catchments and proposing a market mechanism by which heavy water users (forestry, intensive agriculture, particularly sugar production, mining, industries such as paper mills, etc.) can invest in support for sympathetic upland land uses to increase the water yield of the catchments upon which they depend. This market model is finding favour with the South African government as a market means to embed the ecosystems approach as a basis for the equitable, sustainable and efficient provision of water.

PES initiatives are also being developed, for example, in the production of freshwater resources in the drainage basin of Lake Naivasha in Kenya.[34] They are implicit in the intent, if not yet the reality (as discussed in Chapter 2), of the EU CAP agri-environment subsidy system, reflecting biodiversity, characteristic landscapes and other socially valued 'outputs' of land. They feature in novel approaches to flood risk management that step back from merely defending drained land at any cost and instead seek an optimal balance of ecological, social and economic benefits in novel schemes. As noted previously, a 2010 study by the OECD suggests that there were already 300 established PES schemes around the world. PES thereby offers great hope for a more sustainable relationship with biodiversity, not merely subsidizing marginal changes in practice but embedding the value of ecosystem services within the economy. Further potential for market-based approaches to bring ecosystem services into the mainstream is seen in climate regulation, with carbon trading markets now at US state (such as California), national (for example, the UK Climate Change Levy), continental (including the EU Emissions Trading Scheme) and global levels.[35]

What is clear from a range of studies is that effective PES schemes require clear 'providers' and 'buyers' to form a market, for which a strong institution is necessary to form and moderate the market.[36] This is particularly the case given the general need for state and community involvement owing to high transaction costs and the common nature of environmental goods in the world's dominant forms of PES, which are rarely if ever created in an organizational or societal vacuum.[37] Nevertheless, ecosystem services have evolved progressively from useful metaphors and pedagogical methods to become power tools, latterly leading to market and other instruments with considerable promise for addressing many of today's more pressing natural resource problems.[38] They certainly have a key role to play in establishing robust markets to secure the benefits that people derive from ecosystems across living landscapes.

Rethinking the private and public value of catchment land use

Table 7.1 in Chapter 7 reproduces, in modified form, an analysis of the predominant ecosystem service benefits and beneficiaries that may be derived from land and landscapes.[39] Owing to long-established commodity markets, the two provisioning services of 'food' and 'fibre' represent (generally) overwhelmingly private benefit while all other ecosystem services are of more general public benefit across a range of scales. There is, of course, substantial generalization and subjectivity in this assumption, but the aim here is to develop a workable basis for decision- and dialogic-support rather than a definitive decision-making algorithm. Developing this thinking further, a simple model has been published[40] to represent the balance of private and public interests arising from an illustrative set of land uses subject to rapid, semi-quantitative screening using the risk-based screening approach (using the weighting ++, +, O, -, -- or ?) advocated in the Defra's 'An introductory guide to valuing ecosystem services',[41] which is reproduced in Table 7.2 in Chapter 7. The findings of this model of outcomes from land use are reproduced in Figure 8.1. A central assumption is that catchments in a healthy and functioning state generate significant benefits across the categories of ecosystem services. The basic axes of this model of relative private and public benefits are context-sensitive within catchments, but are also value-neutral with respect to current or historic land ownership or agricultural policy. They are rooted in robust principles yet are pragmatic enough to help illustrate, inform and guide decisions in a practical context.

Though illustrative, consideration of land uses under this model provides a helpful and transparent basis for orienting different types of land use against a grid of private versus public benefits, and can also provide a *transparent* basis for negotiation about land use and its subsidy in the context of wider landscapes and different constituencies

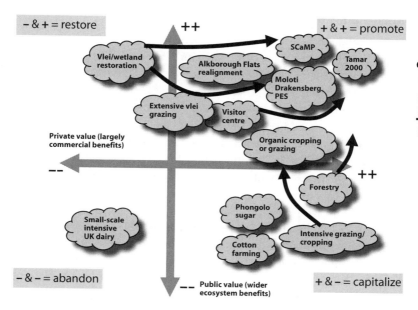

Figure 8.1 Illustrative orientation of land-use options against the axes of private and public benefit (arrows are explained in the following text)

of human beneficiaries. The model overlooks a diversity of stewardship options and arrangements within any one land-use type, but also helps identify how modification of land use could move it 'upwards and rightwards' on the model towards the aspiration of achieving both private and public value simultaneously (as highlighted by the descriptor 'promote' in that quadrant of the model in Figure 8.1). These positive opportunities are discussed at length in the source paper. They include conversion of 'intensive grazing/cropping' to 'organic cropping or grazing' as a means to reduce the overall environmental footprint of land use and restore a range of public benefits from the landscape, and also investment in 'vlei/wetland restoration' to restore the multiple public benefits of degraded wetlands (see arrows in Figure 8.1).

The model thus serves valuable pedagogic, dialogic and decision-support purposes, potentially contributing to wider debates about the sustainable use of land, landscapes and catchments inclusive of the many broader ecosystems and livelihoods within them. Critically, development of this model is informed by, and in turn has informed, consideration of environmental and human development practices with which the source papers' authors were variously engaged in the UK and South Africa.

Applying the value model in the real world

The model of private and public benefits outlined above has been applied in practice in seeking an equitable and sustainable basis for management of, and charging for, water in the Mvoti catchment in KwaZulu-Natal, South Africa.[42] This has stemmed from practical capacity-building work with stakeholder communities across the Mvoti catchment, engaged in a process of dialogue leading to proposals for setting up an institution (a Water User Association) through which the local people propose to negotiate the allocation of water in the Mvoti catchment.[43] An important attribute of the model for this and many other practical purposes is that it does not consider who owns the land, and therefore who may, however misguidedly, assume private rights to the use of the land and its associated water resources. It is focused instead merely upon ecosystem services likely to be produced by high-lighted habitats and land uses, and the balance of private and public benefits accruing from them. Hence, it provides a basis for who should pay for 'production', or be rewarded for 'consumption', of a shared 'common' of catchment ecosystem services. For purely exploitative uses of land that generate private profit while degrading ecosystem services of potential value to broader stakeholders (i.e. 'intensive grazing/cropping' in Figure 8.1), the model's implication is that charges should reflect damage caused and services 'consumed'. However, for less destructive uses of the land that preserve or even enhance ecosystem services of value to broader constituencies (such as SCaMP or the Tamar 2000 outcomes in Figure 8.1), it is appropriate that charges are lighter or even that subsidies should be payable on the basis of hypothecating charges extracted from more damaging uses elsewhere in the catchment. This effectively creates an internal catchment market, not merely in the recirculation of money from destructive uses to more benign or restora-tive uses of land and water, but because the sliding scale of charges itself creates an incentive for land users to modify their management regimes to lighten both environmental impacts and associated charges.

Embedding ecosystem approaches in real markets may be an equit-able mechanism to bind communities together across living landscapes. As water is such a crucial resource not just in the arid landscape of South Africa but elsewhere in the world, river drainage basins (catchments) constitute focal landscape units (in effect a semi-closed system) around which to create these markets. (Water transfer schemes and substantial volumes of 'virtual water' exported in the form of water-intensive crops[44] add complexity but do not prevent the creation of such markets.) Within a catchment system, 'real world' markets may already exist in some services – for example, where stock grazed on uplands is sold to populations in the lower catchment. However, most of the vital ecosystem services provided by catchment habitats are external to current financial markets. The private and public value model (Figure 8.1 derived from the source paper[45]) identifies principal vectors of change towards desirable markets. Stakeholders' views and consensus about the principal habitat types, land uses, human populations and livelihoods, and the functioning of the Mvoti catchment were captured from stakeholder dialogue meetings in the form of maps. Each of the generic land uses identified was subsequently subjected to a rapid peer-reviewed ecosystem service assessment of likely private versus public benefit used to develop the private/public benefit model (see Figure 8.2).

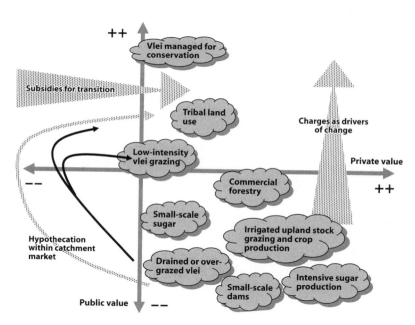

Figure 8.2 Economic forces and connecting markets in a theoretical closed river catchment, and illustrative orientation of land uses reported in the Mvoti against the axes of private and public benefit

Based on the internal market model, there is significant scope to place charges against a range of established profitable land uses that are damaging to the 'public services' of the Mvoti catchment. These include in particular: 'intensive sugar production'; 'small-scale dams'; 'irrigated upland stock grazing and crop production'; and 'drained or over-grazed vlei' (which have the most negative public value in the 'capitalize' quadrant highlighted in Figure 8.1). 'Small-scale sugar' and 'forestry' are assessed as somewhat less publicly damaging, though clearly there is a broad spectrum of potential impact within any one land-use type dependent upon intensity, inputs, practices employed, location relative to sensitive zones within the catchment, etc. According to the market considerations above, all of these commercial uses should pay charges – perhaps levied via water service charges – ramped according to the scale of their impact on the 'public value' of the Mvoti catchment. All land users have the opportunity to invest in best and improving practice, or transitions to other land-use practices, to avoid or minimize these charges throughout a greater net contribution to 'public value'. The proposed charge, if developed proportionately to likely impact on public benefit, could then itself constitute an economic instrument to drive improved environmental practice.

The three remaining land uses recognized by stakeholders in the Mvoti catchment – 'low-intensity vlei grazing', 'tribal land use' and 'vlei managed for conservation' – are approximately neutral in terms of private (commercial) benefit but have different scales of associated public benefit stemming from their impacts on ecosystem services. Arguably, they warrant different degrees of subsidy from funds collected as charges to incentivize and promote the maximization of public benefit.

Where land managers may wish to abandon or radically change land uses to deliver greater public value, there may be a case for hypothecating charges in appropriate and appropriately 'ring-fenced' subsidies. The arrows in Figure 8.2 signify two such options for abandonment of the marginally profitable but ecologically damaging practice of 'drained or over-grazed vlei', taking the land out of production and into the 'restore' quadrant, where subsidies may help the transit of these potentially important wetlands into more publicly beneficial 'low-intensity vlei grazing' or 'vlei managed for conservation' uses.

The private/public benefit model and its application in the 'real world' context of the Mvoti catchment helped stakeholders think about the interactions and interdependencies of their continuing or proposed land uses. Aside from its pedagogic value, it provided an open and transparent dialogic space divorced from issues of land ownership and assumed rights. It served the purpose of an interactive focal point for identification, dialogue and testing of potential solutions to optimize the public value derived from land and water use. It has consequently also helped develop a proposal for an internal market within the catchment to charge more ecologically degrading practices and favour those that better enhance or restore the ecological functions of the catchment from which diverse public (generally non-market) benefits flow. At this point, we should be clear that we are focusing only on internal ecosystem service markets in the Mvoti catchment and are excluding from consideration wider contributions of commercial farming to the economy, food security and employment. However, these economic considerations become pertinent as options for the self-funding of new stakeholder-led catchment management institutions – particularly Water Users Associations (WUAs) and Catchment Management Agencies (CMAs) – required under South Africa's National Water Act 1994 (Act no. 36 of 1998).[46]

As noted above, the 'real world' situation is significantly more complex. Many uses of water are consumptive, such as spray irrigation or paper manufacturing, while others, such as cooling, may return all or some of the water into the catchment, albeit often in a different quality and place. Some water may effectively be 'sold' on to foreign countries embodied in water-intensive crops (rice, cotton, wheat and so forth)[47] or products that require water-intensive manufacturing (such as mined metals). Charging schemes then should be sensitive to water use and contamination in intensive land use and other extractive processes. Likewise, the 'real world' situation in South Africa is one in which historic investment in infrastructure diverts flows from many of the south- and east-flowing rivers in KwaZulu-Natal (KZN) northwards into the economic heartland of the province of Gauteng; abstraction for these purposes is a wholly consumptive use of water which deprives local KZN people of its benefits, and should therefore attract substantial and perhaps punitive charges to safeguard indigenous catchment livelihood opportunities as part of the nation's post-apartheid equity agenda. However, the general principle of taxing environmental 'bads' and recirculating this into incentives for public 'goods' is well established[48] and can be fruitfully applied in the internal market represented by the semi-closed system of a river catchment. This is particularly prescient when stakeholders are required to develop institutions to learn to share water, such as under South Africa's post-apartheid constitution and its far-sighted water laws with their express intention of addressing the three principles of equity, sustainability and efficiency.

Reanimating landscapes

The biophysical reality that humanity is fully dependent upon the services provided by ecosystems may have become buried in societal consciousness under assumptions about private land rights, the annexation of crucial resources such as soil, water, minerals and biodiversity by a privileged minority, market pressures to liquidate common assets for short-term gain, and the global reach of empire-building and its analogue in powerful multinational businesses. Nevertheless, we erode the Earth's supportive ecosystems at our considerable peril, and so need to find novel and sustainable ways of interacting with land and landscapes. The solution is not to seek to arrest the tide of human and economic processes, but rather to use our faculties for systems thinking, foresight and innovation to develop practices and markets that connect the workings of supportive landscapes and ecosystems with their benefits for cultural, social and economic progress. The PES approach, which links 'providers' and 'users' of ecosystem services, has a key role to play here as part of the broader reconnection of cultural perceptions about the importance of land and landscapes and all those who share them.

Prevarication is no longer scientifically, economically, politically or morally defensible if we are to develop the 'living landscapes' – locally, regionally, nationally and globally – that will secure a future for all. Nor do we now lack the basic tools, so there is no justification for continued inertia. We are indeed beginning to grapple with the intricacies of this new market model, and the re-evaluation and reanimation of landscapes, tentatively dipping a toe into the systemic paradigm as we explore the untested waters beyond the assumed safety of inherited assumptions, habits and market pressures. Live efforts to create such ecosystem-based markets in the Mvoti catchment of South Africa, and evolving 'paying for ecosystem services' markets in the USA, Costa Rica, Africa, various European countries, Australia and other nations beyond, show what is possible,[49] while also revealing the complexities and 'hangovers' from our industrial past as we strive for the new paradigm wherein sustainability and equity lie. What is clear is that we know enough about the required steps and the necessary 'direction of travel', and also about the threats and escalating costs of further delays, to take decisive 'no regrets' action now to realize the benefits that this will yield for all people.

9 | Lessons for tomorrow's world

As we approach the latter stages of this book, it is important to 'sweep up' many of the lessons learned, and to think about how they need to shape multiple dimensions of human decision-making as we proceed forward into a future that will be inevitably more populous with needs to be served from a finite and already degraded global landscape.

The positive examples of implementation of ecosystem services to flood risk, fresh water production and river restoration outlined in the preceding chapter represent progressive initiatives that make connections between 'producers' and 'users' of ecosystem services that reinvigorate the economy and the societal valuation of living landscapes. All represent significant strides forward beyond inherited stewardship, governance and economic models, entailing far greater degrees of stakeholder participation and realization of a broader set of interconnected, ecosystem-derived benefits. Importantly, all make progress towards the central importance of safeguarding the diversity and scale of ecosystems to maintain their capacities to support human needs, reflecting the inherently systemic concept of sustainability that addresses the capacity of whole socio-ecological systems to continue not merely within narrow geographical and temporal scales but indefinitely.

These encouraging 'green shoots' of changing attitude and management may constitute 'punctuated equilibria' in the institutionalization of sustainability and the primacy of ecosystem services provided by living landscapes in the decision-making process.[1] However, these early responses in the transition towards systemic ways of thinking and living are as yet far from proportionate to the scale and pace of the threats facing humanity, held back as we are by the inertia of legacy discourse and regulations as well as established governance and economic models.[2] This chapter, then, explores the logical extension of the systems paradigm beyond science and into practice, including conceptual, economic, educational, legislative, institutional and other implications for the fully sustainable management of our relationship with our supportive landscapes.

Various attributes of this transition have been uncovered by prior case studies, and are extended here. These attributes include recognition of ecosystems as critical capital, sublimation of competing disciplines, visionary legislation, inclusive decision-making, representative valuation and markets, institutional and funding reforms, pragmatic tools, visionary planning, novel models of governance, respect for traditional wisdoms, and rights to land, landscapes and their associated ecosystem services.

Recognition of ecosystems as critical capital

Key among assumptions 'hard-wired' into economic and corporate governance systems as well as public consciousness established at the outset of the Industrial Revolution is a reductive view that ecosystems are resources that can be traded and liquidated for commercial gain, and that money can then buy the additional services that we need. This perspective of natural resources as property was further compounded in the holistic era when emerging concerns about the environment improved recognition of environmental impacts, but often framed environmental considerations as a net constraint upon 'development'. For example, one still reads regularly in the media about the irritation of developers at having to take account of organisms scheduled as of nature conservation interest, which are perceived as little more than an extra expense and inconvenience. The net result of these unfortunate perceptions of nature is that ecosystems and natural resources are inherently exploitable or else a constraint on legitimate 'growth', both views combining to create a progressive erosion of the quantity, quality and perception of ecosystems and their capacities to sustain humanity. On a finite planet, with dwindling resources to be shared by a booming human population, we have to challenge and overturn that assumption.

The systemic paradigm encapsulates a wholly different conception of value, founded upon the fundamental and largely irreplaceable importance of ecosystems and the many benefits that they confer upon humanity as core resources securing and advancing human welfare, economic opportunity and future potential. Land, landscapes, water fluxes through them and their associated ecosystems then represent value and opportunity rather than cost and constraint; a very different world-view.

As the Millennium Ecosystem Assessment Board Statement[3] notes, 'In the midst of this unprecedented period of spending Earth's natural bounty, however, it is time to check the accounts. That is what this assessment has done, and it is a sobering statement, with much more red than black on the balance sheet.' The risk of doing nothing is to continue to allow ecosystems to decline in quantity and quality, with dire consequences for the potential for human quality of life. The challenge for all sectors of society is to accelerate progress towards sustainability, with nature conservation and the equitable and sustainable sharing of ecosystems playing a central role in its achievement. This represents an integration of value systems within a sustainable development model that simultaneously recognizes the need for ecological, economic and diverse societal perspectives. Today, progress is hesitant and as yet far removed from the goal of systemic practice, with policies, regulatory processes and practical decisions still tending to revert back to 'win-lose' trade-offs typical of the holistic paradigm rather than the 'win-win-win' innovations that typify sustainable and systemic solutions. Protecting the already strained integrity of our ecosystem inheritance also serves as a moral guide with respect to the safeguarding of nature and its capacities to support human needs and potential, including those of generations as yet unborn.[4]

Language is important here. Reductive simplifications of ecosystems merely as 'resources' and holistic terms such as 'trade-off' or 'mitigation' are often wrongly assumed to talk of sustainable progress but, for reasons already discussed, merely perpetuate misconceptions and misapplication. This matters, as everyday discourse as well as that used in legislation encodes particular world-views that frame public understanding as well as policies, regulatory practice and their acceptance. Other language may be transitional. For example, throughout the 1980s and 1990s there was an increasing tendency in government and its associated regulatory community towards the language of 'partnership', progressing from a narrow framing of 'consultation' (which as we have discussed often entailed opportunities to comment on forgone 'expert' decisions) through to collaboration and subsequently early progression into the systemic paradigm through stakeholder engagement. The systemic paradigm calls for a wholly different basis for decision-making with respect to environmental management, requiring the participation of all interests. Although conflicts between desired construction schemes and other development will doubtless still exist in a sustainable future, the inherent worth of ecosystems as core and often irreplaceable capital needs to become better appreciated and progressively embedded in public policy, legislation and decision-making processes.

Sublimation of competitive disciplines

One of the implications of a systems approach is the need to surrender narrow discipline-specific (reductive or holistic) approaches to environmental management that vie with each other for priority and resources. It is implicit in ecosystems thinking that, as 'emergent properties', the diverse benefits that humanity derives from ecosystems in the shape of ecosystem services have to be considered as an integrated suite. As we have seen, to select and seek to optimize on a discrete service-by-service basis tends to overlook and often degrade other services, to result in 'win-lose' trade-offs, and to perpetuate the degradation of ecosystem integrity and function over broader temporal and geographical scales with implications for human well-being.

It then follows that the narrow disciplinary focus of environmental management is likely to have as risky a set of consequences as narrowly framed commercial exploitation of ecosystems.

We are not talking here of surrendering specialist expertise. Indeed, specialized skills in management of flood risk, nature conservation, fisheries and other discrete environmental disciplines remain of fundamental importance to inform decisions that incrementally deliver a better-connected future. However, we will need to move beyond discipline-specific, localized decision-making (the reductive paradigm) and the 'trading off' of one disciplinary perspective with other perceived competing interests through post hoc mitigation measures (the holistic paradigm) in favour of combining different strands of specialist expertise and other forms of knowledge to innovate novel 'win-win-win' solutions. Examples of these integrated, ecosystem-based approaches to flood risk management, fishery recruitment, water yield and nature conservation are reviewed elsewhere in this book. Further examples of cross-disciplinary solutions are seen in urban planning in terms of the 'ecodesign' of buildings, sustainable drainage systems (SuDS),[5] 'green infrastructure'[6] and greater consideration of the human scale in city planning.[7]

The urgency of the need for this transition is increasingly evident in all environmental disciplines. The surrender of disciplinary identities within governance is then an essential prerequisite of a shift in emphasis from fragmentation towards treating socio-ecological systems in an integrated way. The primacy of ecosystems as life-support mechanisms, and of the diversity of those ecosystems as indicated by rare species and the breadth of genetic heritage, cannot be fully realized by the 'special pleading' of holism, which is more likely to result in localized 'win-lose' trade-offs in exchange for continued ecological degradation across broader rural and urban landscapes. Rather, sustaining habitats and species of societal importance will rest on new approaches and publicly persuasive arguments that reflect the dependence of societal needs, values and aspirations upon diverse and productive ecosystems. The current transition in perspective of flood 'problems' no longer purely as natural phenomena but as largely a product of society[8] needs to be replicated across all other spheres of environmental management to reflect the primacy of functioning ecosystems in supporting human needs on a sustainable basis.

Visionary legislation

The tendency under historic paradigms was for regulatory intervention in environmental management only after gross impacts had become manifest, and then often shaped by associated vested interests. Two British examples exemplify this. The first is the 'great stink' of the River Thames in central London in the hot summer of 1858, when the stench of sewage rising up from the River Thames was so offensive to MPs that Parliament had to be suspended, leading government to agree to 'take immediate measures for abating the dangerous nuisance caused by the noxious state of the Thames' and also ultimately to face up to the need for an urban planning policy more widely in the country. The second example is the 'Great Smog' (or 'Big Smoke') of 1952 London, constituting one of the most significant pollution events in history, resulting in thousands of premature deaths as well as the stimulation of environmental research, government regulation and public awareness of the relationship between air quality and health. Many lesser examples can be found, such as the banning of imports of problematic alien species only when gross impacts have arisen from invasion and perturbation of native ecosystems, when the opportunity for effective control has passed, or else an equivalent naive approach to banning synthetic chemicals only once gross effects have manifested themselves and been proved beyond doubt, yet which have to then be managed in the face of strong vested economic interests. Today, we are waking up to the daunting threat of climate change with various tentative economic measures being initiated in the face of a global industrial model largely founded on investment in petrochemical energy. Clearly, legislation that merely reacts, generally slowly and ineffectually, to the worst impacts of substances, products, processes, introduced organisms and land uses is at best a metaphorical 'sticking plaster on a wound' and, at worst, a positive inducement to continued reductive commercial exploitation of fundamental assets blind to longer-term consequences. This has led some to argue that the proliferation of environmental treaties and other international instruments has failed adequately to address the problem of environmental degradation, demanding a new approach to international law with an ecological basis.[9] Other commentators are more damning of the pace and reach of statute law, some arguing that the pace at which it develops means that statute law is almost automatically out of date and/or too much influenced by vested interests by the time it is enacted.[10]

This is perhaps not a fault of statute law itself, but of the paradigms under which much of our legacy of environmental legislation was framed. We are now better equipped to take bolder and better-understood steps forward, and the right sort of legislation remains a vital element of this. Statutory backstops provide an important 'bottom line' of behaviours based on culturally shared agreements. The transition into a systemic management paradigm is at present preliminary, albeit growing in momentum.

In India, for example, the Forest Rights Act (India) 2006[11] was enacted to address concerns about the rights of forest-dwelling communities to land and other resources which had been denied to them owing to the perpetuation of colonial forest laws. The Act includes simultaneous redress of historic injustices to tribal forest-dwelling people and conservation both of nature and of the natural resources underpinning the livelihood needs of these people. Tribal people dependent on India's forests for harvesting of produce and water, grazing and shifting cultivation are counted in their millions. The Act has its critics, including those who disagree with the passing of ownership to formerly landless people and allocation of right of access to biodiversity, intellectual property and traditional knowledge, and including some environmentalists who fear for nature conservation interests. Nevertheless, the Act has now passed into law and its intention to integrate indigenous rights and the long-term protection of forest biodiversity and ecosystem services supporting the livelihoods of many people living outside of or peripheral to the mainstream economy is inherently systemic in its intent.

In the UK, the EU and South Africa, various legislative strands begin to integrate management of land, water and their human use as an integrated whole. For example, Defra's *Making Space for Water* strategy and the Environment Agency's emerging *Flood and Coastal Risk Management Strategy 2009–2015* (both considered in prior chapters of this book) indicate the emergence of strategic thinking about flood risk management significantly in advance of current institutional arrangements which, by and large, remain fixed in the holistic model. The systemic vision inherent in the EU Water Framework Directive (WFD) and South Africa's National Water Act 1998 (NWA) both illustrates legislation based on systemic principles and also serves to exemplify the far from seamless, transparent or rapid translation into everyday operational practice. There are, in both instances, significant lags between perception and thinking, ensuing policy and eventual management practice. The Water Framework Directive, for example, was gazetted by the EU in 2000, transposed to the national scale, incorporated into policy by national institutions, and was at every stage influenced by established habits, assumptions and limitations of monitoring data, not only instituting delays but also representing potential opportunities for misinterpretation and/or weakening of the clear systemic aspirations of the WFD.[12] Similar practical difficulties are seen in the extremely slow pace of implementation of South Africa's NWA, the full systemic vision of which, now well over a decade later, has yet to be attained owing to a range of institutional, capacity, financial and technological barriers.[13] The integrated economic, social and environmental approach to whole catchments that the NWA and WFD embody is broadly consistent with the intentions of Integrated Water Resources Management (IWRM),[14] an inherently systemic management framework evolving since the early 1990s. IWRM addresses the integrity and functioning of catchment ecosystems and their human uses and modifications as integrated systems, addressing hydrology (with its implications for flood risk and water resource provision), protection and enhancement of habitats for nature conservation, carbon sequestration, diffuse pollution control, fishery vitality, physico-chemical processes influencing water quality, and cultural uses and appreciation of catchment landscapes. As with the EU's WFD and South Africa's NWA, realization of the promise of IWRM remains patchy and partial, reflecting how even the boldest, clearest and most consensual intentions for society to move to a fully integrated, systemic basis for management of complex systems are far from complete.

Other positive examples of systems-based legislation include the import of fish into the UK under the Import of Live Fish (England and Wales) Act 1980 (ILFA) and the Import of Live Fish (Scotland) Act 1978, both of which take a precautionary approach entirely consistent with systemic management. This is achieved by banning, as a matter of principle, the import and potential introduction of alien fish species considered capable of forming self-sustaining populations in British waters. By contrast, the import of plants into the UK is governed still by legislation that reacts, often over periods of many years, on a species-by-species basis only when there is evidence of serious adverse ecological impacts. By this time, establishment will have occurred and eradication will have become all but impossible, while import and trade continue largely unabated for plant species not yet deemed problematic or else not having completed the long and contested process of scheduling as such in legislation. By contrast with the approach to alien fish, UK regulations covering alien plants seem still to be framed by the paradigm of holism, founded on discrete reactive responses to individual 'problem' species while failing to constrain the lucrative horticultural and agricultural trades which often underpin emerging problems; the consequent legislative emphasis is upon plant diseases (e.g. the Plant Health Act 1967 and its many amendments) rather than on principles related to an inherent potential for naturalization of alien species with all of their unforeseeable and often adverse and costly consequences. We can add to the list of positive legislative transitions the intent of EU REACH legislation for control of bulk chemicals, which focuses on inherent properties of substances – potential for bioaccumulation, persistence, reprotoxicity, carcinogenicity, and so forth – in preference to former approaches to banning chemicals on a substance-by-substance basis only when such symptoms manifested.

The key message emerging here is that visionary legislation that focuses on principles rather than reacts to impacts will be essential to guide us strategically towards sustainable goals. Today, systems and ecosystems approaches are therefore beginning to influence some of the world's more progressive legislation, although the extent and pace of uptake of these progressive ways of thinking into the mainstream of public policy need to increase dramatically if we are to break away from the kinds of blinkered assumptions perpetuating today's pressing problems.

Inclusive decision-making

Much inherited environmental regulation and decision-making is based on the holistic assumption that conflicts between 'silos' of human interest can be managed by mitigation of unintended impacts largely retrospective to the primary decision. This approach favours decision-making by narrow interests instead of promoting participation in initial decision-making involving broader networks of stakeholders sharing the fate of common natural resources. There is then a need for inclusive methods of decision-making that take a systemic approach to all likely impacts of proposed products, processes, land uses and other practices for both ecosystems and their many beneficiaries. This is not only a recipe for more equitable outcomes, but can also better ensure that all of the diverse consequences for ecosystems are factored into the decision-making process via their multiple beneficiaries. A practical example lies in different land-use options for flood-plain farmland which have different degrees of compatibility with a range of other ecosystem-mediated benefits, often with surprising interactions, such as the improved outcomes for natural flood regulation processes arising from intensive grazing as compared to the ostensibly more 'natural' management of water levels for the conservation of water birds.[15] All land, water and other resource-use decisions have ramifications for broad constituencies of stakeholders often historically overlooked in decision-making yet all with a stake in the outcomes of management decisions. All are also volatile with, for example, fluctuations in farmed commodity prices or public attitudes to nature conservation, demonstrating the need for decision-making that is not only dynamic but also adaptive.

In addition to conceptual, legislative and economic progress, it is vital that governance systems evolve towards a more inclusive approach recognizing the validity of all voices within complex socio-ecological systems. Contrasting the DAD model of decision-making (decide-announce-defend, as described in Chapter 4), an artefact of our reductive and holistic past but also sadly still prevalent as we move into the systems paradigm, Lindsey Colbourne[16] highlights the importance for flood risk management of moving towards a model of 'engage-deliberate-decide', or EDD. This observation is as relevant to the management of all aspects of the environment. Among the key features of the EDD model is that considerable effort is spent initially in engaging with and understanding the perspectives of different stakeholders at the outset when problems are framed. In contrast to what happens with the DAD model, 'experts' (including discipline-focused specialists as already outlined) support rather than dictate the EDD process which concludes, rather than starts, with decisions. There are indeed already legal instruments requiring this change in dialogue, including for example the international Aarhus Convention 1998 (introduced in Chapter 2), which demands public engagement in environmental decision-making leading to co-creation and co-ownership of solutions right from the initial problem-framing phase through to options identification, options appraisal, decision-making, implementation and lifetime management. The requirements of the Aarhus Convention have already been embedded in the drafting of the EU Water Framework Directive, but also apply retrospectively to all pre-existing environmental legislation. Stakeholders representative of affected communities should be engaged from the earliest phase and throughout scheme life. There is no guarantee in this process that a problem perceived by experts as relating to, for example, flooding will be defined as such by the stakeholder group, opening the way to far greater vision of societal perspectives on environmental issues that affect them and hence the potential for more broadly framed, innovative solutions. Importantly, at the points at which decisions are made, stakeholders will understand and feel engaged, and better able to support them even if they do not fully meet the individual wishes of every stakeholder.

A further important aspect of this approach is that it is an inherently learning approach, adaptive to the volatility of the 'real world' and taking account of multiple uncertainties, such as a changing climate, unpredictable or non-linear ecosystem responses, livelihood transitions, and external pressures comprising both continuous trends and discrete 'events' (for example, major floods, sudden energy price hikes, climate change trends and thresholds, or novel crop or stock diseases).

This progression towards public participation is not unknown in environmental governance in the UK. For example, the democratic intentions of the post-Second World War planning system were clear in the legislation. However, the stakeholder approach has been slow to percolate into management of the wider environment, perhaps because of former extensive domination by 'expert' engineering perspectives. Nevertheless, we are beginning to see increasing partnership-based and dialogic approaches, particularly in relation to water management. These include emerging UK approaches such as Water Cycle Strategies, which are intended to integrate aspects of water management in the built environment, including waste water, WFD considerations and flooding from surface, fluvial and tidal sources. Further UK water-based partnership approaches include consultation on the Catchment Flood Management Plan for the Thames,[17] which looked at the entire catchment over a planning horizon of fifty to one hundred years to help break out of fixed assumptions about urban development and land uses. The Environment Agency's consultation draft of the *Flood and Coastal Risk Management Strategy 2009–15*[18] takes a similar long-term and partnership-based approach. These initiatives also span broader disciplines by seeking to make greater explicit contributions to wider government agendas for economic growth, the environment, social justice, working with natural processes, land management, climate change considerations, and a commitment to collaboration with partners.[19]

It is clear that the transition to embracing wider perspectives of the environment and the many ways in which all stakeholders affect it, and are in turn influenced by it, is at least recognized. It will be necessary to engage stakeholders strategically and at greater pace in deliberative processes if we are to accelerate systemic transition towards sustainability.

Institutional and funding reforms

As covered in considerable detail in Chapter 7, the real costs and benefits for ecosystems and their many and diverse beneficiaries are currently poorly reflected in all development decisions. There is a pressing need to develop markets to account for the full range of impacts, positive and negative, that our various activities impose upon the ecosystems that underwrite human security, economic opportunity and future potential. This must also include broader societal perspectives and value systems. Developing the market not by putting 'prices on nature' but by reflecting the many different values bestowed upon all in society by ecosystems is a potent means to advance equity and sustainability. It would achieve this by broadening awareness and appreciation of these broader values, creating incentives for sustainable innovation, and embedding the full ramifications, benefits but also liabilities in decision-making. This may represent a more pervasive means of effecting change than can be achieved through legislation.

However, narrowly defined management budgets today tend to enforce reductive fragmentation in ecosystem management, inhibiting the wider framing of issues and solutions. More widely defined budgets are therefore required to reflect both the breadth of stakeholder concerns and the opportunity for multidisciplinary solutions. It may be necessary to retain some core funding to reflect disciplinary perspectives, at least in a transitional phase, as for example to support 'hard' flood defences protecting publicly valued landscapes, infrastructure and built heritage that is not best located from an ecosystems perspective. A practical example of this is the need to 'defend' heritage and critical infrastructure not sited ideally with respect to river corridor hydrology, constraining the more ecosystem-centred aspirations of the 'Blue Corridor Vision' to alleviate flooding from the River Derwent in the English city of Derby.[20]

South Africa's visionary National Water Act of 1998 is again a leader here. It contains explicit provision for the establishment of new, more appropriate stakeholder-based institutions to promote the sharing of water, simultaneously addressing the three driving principles of equity, sustainability and efficiency. This explicitly recognizes that the institutions instigated under the former political regime also had narrowly framed duties and budgets that may cement yesterday's assumptions. Institutional realignment – the reform of institutions along with their decision-making frameworks and associated societal outreach and participation – is an essential element for promoting and maintaining transitions in management, community participation and local decision-making. This is also implicit in Europe under the WFD, although progress to reform the structure and/or funding of institutions to achieve 'good ecological status' is not thus far proceeding rapidly. The key principle here is that narrowly 'ring-fenced' budgets tend to drive narrow disciplinary decisions that may work against integrated solutions.

Pragmatic tools

There remains a need for the development of pragmatic operational tools to guide day-to-day decision-making on an increasingly systemic basis. They also need to support novel forms of participation and negotiation around the sharing of common resources.

One operational example of such a practical tool is the *eThekwini Catchments: A Strategic Tool for Management*.[21] The municipality of eThekwini, on the coast of KwaZulu-Natal in South Africa, recognized that overuse of river corridors converging on the city of Durban was likely to limit further urban, social and economic development through impacts on the various ecosystem services they provide. Consequently, the municipality commissioned the *eThekwini Catchments* study to explore the status of ecosystem services provided by these rivers. The resulting study went well beyond the basic brief, presenting this information graphically as a red/orange/green 'traffic light' coding representing the state of a range of ecosystem services – air quality, water quality, water quantity, flood risk, sedimentation/erosion, loss of biodiversity, agricultural production, and recreational/cultural/educational uses – for each of the tributary rivers. This was structured as a practical guide with direct applicability to day-to-day planning, supporting the decisions of planning staff on a fully transparent and readily understood basis. For example, if a development proposal with a large impermeable surface were to be permitted in a Durban sub-catchment already coded 'red' with respect to flood risk, the developer is graphically presented with the knowledge that exceeding this aspect of ecosystem carrying capacity can only hike flood risk elsewhere in the city. This knowledge is also visible to other stakeholders in the decision. By incorporating ecosystem services in the tool in ways that connect the capacity of ecosystems, consequences for the best interests of diverse urban stakeholders, and associated economic implications, a systemic perspective is instituted in publicly accessible planning processes. The *eThekwini Catchments* planning guide has been used routinely since publication in 2002, and has been instrumental in informing a wide range of decisions both routine and controversial. It also provides development proponents with an indicator of the likely obstacles they will encounter from the perspectives of both planning authorizations and infrastructure design, streamlining the application process but also helping them innovate novel development and design options to take account of ecosystem service capacity.

Visionary planning

If a whole-systems view of catchment integrity could be conceptualized, various formerly disparate land-use, management and development activities could be better designed and integrated as 'building blocks' towards increasingly intact and functional catchments.[22] These 'building blocks' could comprise, for example, the improved targeting of agri-environmental subsidies better to protect or rebuild critical catchment processes, the strategic implementation of sustainable drainage systems (SuDS) simultaneously to protect water quality, hydrology and habitat in urban development, river restoration schemes designed with their contribution to catchment-scale functioning in mind, or a more sensitive and strategic approach to fishery management. Viewed from the context of a whole-catchment vision, catchment-scale sustainability can be progressively built by investment in discrete yet systemically informed management interventions constituting 'jigsaw pieces' that rebuild ecosystem processes.[23]

As addressed previously, there is also a need to bridge the perceived gap between the historic focus on rare species and habitats under the former model of nature conservation and the broader set of ecosystem services of which they are part. The 'Lawton Review' of nature conservation in England[24] highlights that conservation measures can co-deliver wider ecosystem service benefits, that the conservation efforts need to break out from the holistic emphasis on preserving wildlife sites to enable species to move in the wider countryside in order to adapt to changing environmental conditions, and that far greater emphasis has to be put on sympathetic management at landscape scale to protect both species and the many other ecosystem services from which society benefits. This perspective is substantially endorsed by a wider consensus that it is essential to keep biodiversity viable outside of conservation and land both for nature conservation and ecosystem service purposes, with an illustrative focus not on saving 10–30 per cent of tropical biodiversity in 1–2 per cent of the land, but saving 80–90 per cent on 5–15 per cent of the land.[25] The ways in which ecosystems 'produce' many of these services are far from fully understood,[26] particularly so for 'supporting services' essential for maintaining ecosystem integrity and functioning. Therefore safeguarding these irreplaceable life support services demands of us a precautionary approach that asserts there should be no net loss of either the diversity or quantity of nature as a means to protect their resilience and future value. Nature conservation initiatives are thus readily brought into a more systemic approach to land-use planning.

The slow but steady transition towards systemic approaches to flood risk management provides opportunities for visionary multi-benefit planning. So too does improved targeting of agri-environmental payments for the maximization of public benefit through protection or enhancement of ecosystem services, as discussed in preceding chapters. Elements of freshwater fishery management interests have also been among the leaders in whole-catchment restoration, with a changing emphasis on more ecosystem-centred approaches to environmental management, recognizing the desirability, economic value and conservation needs of self-sustaining fish stocks.[27] This includes in particular the promotion of naturally spawned salmonid species championed in the UK by organizations such as the Wild Trout Trust,[28] the Salmon and Trout Association[29] and the river trust movement (including its umbrella body, the Association of Rivers Trusts[30]). This can only be achieved by working to overcome 'bottlenecks' to the natural recruitment and production of native fish in rivers, requiring a change in paradigm of fisheries management that emphasizes restoration of river habitat, activism and targeted farm business advisory visits to address water quality concerns, rehabilitating natural hydrological regimes and intact ecosystems.[31] As Cowx and Collares-Periera[32] note, '... if conducted in a comprehensive manner, involving the wider public and all stakeholders, fish conservation management will confer wider environmental benefits and protect biodiversity for future generations'.

Indeed, some larger and more culturally valued migratory fishes may serve as 'iconic species'.[33] These 'iconic' fishes are generally top predators moving through networks of connected habitats to complete their migratory life cycles, and so indicators of ecosystems in intact, connected and functional states, potentially focusing public support, spiritual and other values in much the same way as 'flagship' terrestrial conservation species such as tigers and elephants serve to help safeguard networks of linked habitats in India. However, these 'iconic' fish species also have exploitative values through recreational angling, commercial and artisanal fishing, as well as other dimensions of cultural importance. They can thereby form a focal point for all sectors of society for the integration of management actions necessary to protect or rebuild the integrity and connectivity of broad networks of habitats required across whole life cycles, and encompassing human pressures ranging from land use, impoundments, impacts of stock animals, pollution in its various forms, habitat change, species introductions, and positive river and catchment management. Successes with localized conservation of Indian mahseer fishes (of the genus *Tor*), Atlantic salmon (*Salmo salar*), American alligator gar (*Atractosteus spatula*) and paddlefish or 'spoonbill' (*Polyodon spathula*), and southern Africa's yellowfish (species of the genus *Labeobarbus*) demonstrate the potential of these 'iconic species' to mobilize wider public concern and investment in protection and restoration of sustainable populations, with all the associated benefits that enhanced freshwater ecosystems provide. A prominent example here is the constitution in 1986 of the Thames Salmon Trust, reconstituted in 2005 as the Thames Rivers Restoration Trust, as a registered charity with the ambitious aim of bringing about regeneration of the river such that salmon would again be able to return. Significant press, institutional and public support surrounded the establishment of this trust despite salmon being absent from the river at the time, as the return of this once common, characteristic and charismatic fish suggests to a lay public that the recovery of the formerly highly abused Thames system has been achieved. Trust activities included improved coordination of the limited investments of statutory, local authority and voluntary bodies with a focus on the integrity of the river ecosystem, as well as investment in fish ladders and fish passes to enable returning salmon and migratory trout to navigate obstructions. Continued pressure was brought to bear on the improvement of water quality, and some stocking of juvenile salmon was undertaken. However, the majority of the measures implemented were to lay the ground for the natural return of Atlantic salmon and migratory trout to the Thames catchment. Salmon have indeed since returned to the Thames, albeit in tiny and as yet far from self-sustaining numbers, and today are an indication and an 'icon' of the vitality of the river, with all of its associated benefits not merely to angling and nature conservation interests but also the multiple beneficiaries of the full breadth of associated ecosystem services provided by a river and its catchment in a more naturally functional state.

Indeed, rivers and their wider ecosystems have often provided effective foci for regional regeneration. Examples from the UK include the Mersey Basin Campaign, established in 1985 as a twenty-five-year project to reverse the degraded ecological, social and economic former industrialized heartland of the north-west of England through which the Mersey river flows.[34] Many other river-centred regeneration schemes have taken place across the UK, ranging from South London's River Quaggy, Manchester's Salford Quays, Kent's River Medway, to a network of river systems across the nation often championed by voluntary river trusts. Many of these are reviewed in the excellent book *Urban Rivers: Our Inheritance and Future*.[35] All of these river management measures focus on the integrity and optimal functioning of ecosystem processes, increasingly targeted at the role of river restoration in wider nature conservation benefits, regional economic regeneration and a wide range of other societal interests, which will have inevitable co-benefits for many other constituencies throughout river catchments whose interests are supported by ecosystem services.

Realization of a wide range of public benefits from river restoration has become a truly global phenomenon, including, for example, the Cheong Gye Cheon river restoration in Seoul, the capital city of South Korea. When it was covered over by concrete in the 1930s, the Cheong Gye Cheon was perceived as a threat to the city, polluted and surrounded by slums which flooded frequently. By the 1960s, it had been covered by a highway and effectively lost completely to the community. However, various enlightened decisions were taken from the 1990s, resulting in the river and riparian habitat being progressively opened up and restored in a process of urban regeneration that has recovered historical and cultural value, provided floodwater protection, and seen the return of numerous fish, birds and insect species.[36] The Cheong Gye Cheon river has become a focus for urban regeneration in the city and has also been one of the country's major visitor attractions since its formal opening in 2005.[37]

The key point is that planning has to be undertaken on a visionary basis that focuses not merely on immediate pressures, as much of it has historically, but on broader scales of space and time, including the aspirations of all people sharing common ecosystem resources. Aside from the examples provided above, the consultation draft of the *Catchment Flood Management Plan* for the Thames,[38] referred to previously, illustrates how both looking at the whole-catchment scale as well as taking a fifty-to-one-hundred-year planning horizon can help break out of fixed assumptions about current urban development and land uses. Ultimately, a basis in ecosystem services not only helps communities within catchments and other shared resources identify how they depend upon and interact with each other via their impacts on natural resources, but also how they can determine the kinds of ecosystems they require in future to provide the services enabling all stakeholders better to meet their livelihood aspirations.[39]

Novel models of governance

As society grew in numbers and differentiation, it created institutions to address issues requiring specialist knowledge and consideration.

Today, both the developed and developing worlds are well supported by diverse government departments and their associated agencies, as well as numerous manufacturing and retail companies, 'watchdogs', trade associations, NGOs, research bodies, financial institutions and diverse other specialized institutions. The driving legislation, mission, objectives, finance streams and structure of each bear the legacy of the paradigms and assumptions that formed them. Many developed-world institutions and their budgetary arrangements, for example, remain fixed in disciplinary 'silos' (generally with equally siloed budgets) that tend to cement old, divisive and inherently unsustainable paradigms. It is therefore essential that we assess how 'fit for purpose' our institutions are for driving progress towards sustainability and equity, realigning or reforming them as need be to better orient ourselves towards a more connected, systemic and sustainable future. As a general principle, a healthy civil society reflected by a diversity of engaged NGOs is good for democracy.[40]

In the old top-down model, leadership came from government and its departments and agencies, regulating the narrower self-interested tendencies of 'lower-tier' institutions. As we have seen, this model is not only disintegrating with the retreat from holism but is also flawed to the extent to which the interests of privileged sectors of society were often enshrined in the decision-making of such top-down institutions. This is exemplified widely throughout the European Industrial Revolution and during South Africa's apartheid era, when land- and resource-owning interests generally made narrowly defined decisions based on personal advantage blind to or unconcerned about consequences for other people and ecosystems over broader geographical and temporal scales. This was taken up into regulatory systems, such as resource licensing, balancing conflicts as they manifested within a top-down government model often reflecting the views of advantaged sectors of society, and also by subsidy systems founded on 'profits forgone' (as critiqued in Chapter 2) for the implicit protection of the absolute rights of landowners to act for personal advantage.

Today, we are witnessing leadership oriented towards more systemic and sustainable practice coming from other sectors of society. 'Real world' examples here include the visionary and systemic SCaMP initiative driven substantially by the private company United Utilities, fortuitously placed as owner of upland water capture areas with a vested interest in their more sustainable use as a novel approach to serving its predominantly lowland water service customers on the basis of safeguarding and retaining ecosystem services. Various other

UK water companies, significantly including Wessex Water,[41] have also been innovators of novel ecosystem-based market models to serve customer needs on a more sustainable basis, also going well beyond the traditional expectation of following government regulator compulsions to address environmental and social responsibility. The passing of the era of such exclusively top-down leadership is also seen in the innovation of the New York City water supply system by water service providers, the formation of new, independently certified markets under the Forest Stewardship Council and Marine Stewardship Council schemes driven by commercial companies as well as NGOs and local affected communities, and companies such as the Cooperative Bank, Interface and other leaders driving sustainability reporting initiatives.[42]

Commercial and voluntary freshwater fishery interests and associations have also been prominent in advancing both common and statute law. For example, freshwater fishery interests have been prominent in the safeguarding of fishery ecosystems through advancement of the common law,[43] building up a rich case law of injunctions, damages and other forms of legal action associated with the attribution of 'property' both to fish in enclosed waters and fishing rights and enjoyment of the angling experience more generally in fresh waters. A 'landmark' common law case taken by UK NGO the Anglers' Cooperative Association (ACA) was *The Pride of Derby Angling Association Ltd* v. *British Celanese* (1953), which established damage to the fishery due to industrial pollution subsequently resulting in the courts also requiring substantial investment in municipal sewage treatment infrastructure. Many novel cases have subsequently been brought and won by the ACA (reconstituted in 2009 as Fish Legal), often establishing legal precedents and posing threats to would-be polluters and thereby acting as a de facto regulator.

The old model of top-down government leadership to drive the performance of laggards and integrate sectoral interests across other perceived 'lower tiers' of society is passing in the transition to a systemic and sustainable world. All sectors of society have an equal contribution to make as de facto 'regulators', a far greater role as innovators, and an active stake in governance. We are witnessing some transitions in governments around the world to reflect this. Under the UK's 2010 Conservative–Liberal Democrat coalition government, there is for example an explicit drive towards a 'Big Society' (and a small state) to devolve decision-making to citizens, and of course there is already a Mayor of London and quasi-autonomous governments in both Scotland and Wales. On a more impressive scale are ongoing reforms in Australia under the Catchment Management Authorities Act 2003 to include the voices of stakeholders in local-scale decision-making institutions, some of which is being emulated in implementation of South Africa's National Water Act 1998, which has been discussed previously.

The logical culmination of these trends to devolution is a far more vertically as well as horizontally integrated model of governance that is both participatory, involving stakeholders at all tiers in decisions pertaining to their access to resources and other services, but also adaptive, such that governance and decision-making are more responsive to changes in resource status, societal views and external pressures such as the impacts of climate change. We clearly have a long way to go to break free from yesterday's government-dominated model, opening the door to more engaged governance reflecting the livelihood aspirations of all of society in ways that are not recaptured by powerful vested interests. Yet there is real evidence of the transition in process, and of some government-led initiatives to facilitate it at least in some parts of the world.

Respect for traditional wisdoms

A logical extension of this is to recognize that the wisdoms of people dependent most directly upon ecosystem services, without the support of modern and largely unsustainable energy- and technology-intense lifestyles, not only need to be respected but also learned from as we seek novel means for living in closer balance with the Earth's supportive capacities. These powerless people, often marginalized today by the centralization of influence and resources by advantaged sectors of society, deserve respect for their livelihood needs and aspirations. Inherited traditional subsistence livelihoods may also frequently embody wisdoms and place-based adaptations evolved over millennia to match the carrying capacity of ecosystems. It is in no way irrelevant just because much of it may be external to the world's dominant capitalist economy. As we have seen, many tribal and other communities sharing common land have developed traditions and customs not only reflecting the carrying capacity of habitats but also the social infrastructure necessary to overcome the tendency of 'the tragedy of the commons'. Similar observations have been made about the inherent wisdom of widely abandoned forms of land use such as water meadows in the south of England, some of which have persisted for over four hundred years by tapping into natural flows of energy, nutrients and moisture, as well as natural weed control functions, and thereby representing an inherently sustainable form of agriculture, albeit one that is rendered substantially non-viable by modern energy, tillage and chemically intense land-use practices, but above all by contemporary commercial pressures on farming.[44] In all cases, principles underlying these land-use practices may hold wisdoms invaluable for the evolution of increasingly sustainable land-use practices as energy prices, nutrient scarcities and impacts, and ongoing biodiversity loss make more sustainable forms of food production essential.

Rights to land, landscapes and their associated ecosystem services

Up until the beginning of the nineteenth century in the UK and much of Europe, rights to the use of land and landscapes, watercourses and their ecosystems and natural resources were overwhelmingly invested with landowners, with consequences for others sharing the services of those resources almost completely overlooked. However, as we have seen throughout this book, recognition and progressive safeguarding of the public value of land, water and ecosystem services expanded significantly throughout the twentieth century. Extrapolating this trend into the future suggests a continuing transition from private rights tied to land ownership towards communal ownership of, and decision-making about, the ecosystem services produced by land and landscapes, and ultimately their fair and sustainable use and allocation to secure a sustainable future.

This mirrors the transition from rights-based to interest-based allocation of water under South Africa's National Water Act 1998 as a mechanism for the societal transition from apartheid-era hegemony and privilege towards democracy achieved substantially through access to critical natural resources, including water. While the timescale is likely to be less compressed than that seen recently in South Africa, the legal, economic and broader societal implications of this transition to ecosystems-based governance are at least as great as the intended reforms in post-apartheid South Africa. They also need, urgently, to be spread globally on a concerted basis.

While the goal is a just and sustainable basis for governance founded upon the services provided by ecosystems, rather than their annexation by landowning privilege, there remains a necessity to operate workable land and resource stewardship models that allow the operation of ecosystem-centred markets. There must, for example, be a viable business model for land managers to farm in ways that produce food, water resources, biodiversity, flood risk and water quality benefits, and a balance of other desirable and locally appropriate ecosystem services, in ways that are appealing and offer realistic remuneration. The market itself therefore needs to be reformed to increasingly internalize eco-system services, rewarding land uses delivering public value.

This market-based approach to the sharing of and investment in ecosystem services is not incompatible with established private land-owning models, but does necessitate considerable innovation from today's starting point. Fortunately, as we have seen, we are already in the middle of a significant transition from private to public rights, with associated subsidies and the emergence of some markets and common-law precedents, that reaches back perhaps a century in the developed world. This is an exciting time as global society reaches a threshold beyond which it may step towards an assured future, and certainly a future that will differ radically in character and prognosis from the consequences of continued inaction.

Striving for tomorrow's world

The various dimensions of the transition to a systems-based world reflecting the equitable sharing of valued and essential landscapes can now be clearly defined. We are making considerable early progress with some of these dimensions, while others require of us more work. The need for urgent and concerted transition across society to safeguard the long-term interests of all is becoming increasingly evident and pressing. The courage to accelerate this transition, to reach for a more secure and just future but also to avert the looming crisis of inaction, is something that we must muster and apply with no further delay, putting into action what we now know.

10 | The people's land

We are inescapably born of and fully dependent upon this one precious Earth. We are constructed of its matter and evolved as integral elements of its ecosystems. Finding our place with the land that bears us is crucial for many dimensions of our well-being, and provides us with principles essential for securing our long-term prospects at this turning point in the evolution of our species and the world we inhabit.

The uses, abuses and conflicts that have scarred the face of the Earth and potentially blighted our future are not, as yet, wholly catastrophic. The core resource of land and landscape that supported the evolution of our species has remained constant in its inherent properties and support of our changing needs and aspirations and, albeit to a degraded extent, provides the bedrock of our future potential.

It is then hardly surprising that the Earth has been imbued with spiritual significance throughout the history and geographic span of humanity, ranging from Mother Earth to the Empress of the tarot deck and various other 'Earth Goddess' icons from the Greek 'Gaia' to the Thai 'Phra Mae Thorani' and myriad other equally globally pervasive deities of the soil, land and its fertility. While we would be foolhardy to attempt to forge consensus and cross-societal policies upon this fragmented spectrum of deities, we would be even more foolish to dismiss a pervasive underlying sense of spiritual connection to the Earth that is archetypal to all cultures.

Cultural services

The Millennium Ecosystem Assessment classification of the services provided to humanity by the Earth's ecosystems is explicit about 'spiritual value', 'inspiration of art, folklore, architecture, etc.' and 'social relations (e.g. fishing, grazing or cropping communities)' as significant cultural services derived from ecosystems. In our generally more secular Western culture, we know these manifest in terms of our general sense of place and of wishing to die and perhaps be interred in 'home soil'. These are captured in part by the services noted above but also in the additional MA cultural services of 'cultural heritage', 'recreation and tourism' and 'aesthetic value'. This all represents an explicit intent within the MA classification of ecosystem services to integrate different cultural value systems into a model allowing their inclusion within overall governance of sustainability.

There is indeed considerable cultural relativism in the ways in which different societies conceive and value their connection with the land. In South Africa, for example, my Zulu friends remind me that, in the West, we buy, sell and otherwise ascribe monetary values to land, while their ancestors lived in the landscapes of their homelands of which they themselves are a part. This ancestor-based connection to the land can be found in diverse other cultures from China to aboriginal Australia and many Native American and Canadian First Nation tribes. Likewise, in the Shinto religion of Japan, land and diverse attributes of the landscape are ascribed inherent '*kami*', or 'spirit', that may change with the seasons. This is mirrored in various other animist beliefs across the globe, including, for example, 'Brahmin', contained by all of creation under the Hindu tradition. The various regional assessments of the Millennium Ecosystem Assessment (MA) reveal a fascinating cultural landscape of diverging world-views and values that different societies derive from landscapes and their ecosystems. For example, the MA regional assessments for southern Africa[1] or Indonesia[2] each represent strikingly different spiritual and cultural perspectives compared to those for regions such as Europe, North America and western China which are more deeply founded on the capitalist creed. Through its inclusive framing and further elaboration of the four different categories of ecosystem services – provisioning, regulatory, cultural and supporting – the MA framework of ecosystem services contributes to the healing of disconnections or rifts between different cultural world-views and value systems, thereby potentially reconnecting a range of formerly disjointed value systems and ideologies.

The Earth and human well-being

Various scientists and philosophers have sought to articulate the importance of this connection to land, landscapes and associated ecosystems in a Western context. The natural world is indeed increasingly becoming appreciated from a social value perspective. This includes a spectrum of benefits ranging from support for basic human health and well-being through to the value of urban green spaces for social cohesion, inclusion, tranquillity and inspiration, and the presence of and access to nature as major contributors to regional distinctiveness and quality of life.

At its most basic, this entails moderating our impacts on the environment to prevent unintended human health impacts. Early air and water pollution control legislation, for example, was generally triggered by public health rather than inherently environmental concerns, and health-related issues, particularly those affecting children and other vulnerable groups, are widely recognized as 'shock factors' likely to trigger public demand for action.[3] Often, the health or aesthetic implications of the environment are most keenly felt at local scale, or as they impinge upon closer relatives and local communities.

In parallel with this, there has been a growing movement towards wider recognition of the generally non-economic, social valuation of the environment for physical, psychological and spiritual health. Some significant breakthroughs in thinking are reflected in the books of E. O. Wilson,[4] Ted Roszak[5] and others, which progress thinking about the deeper implications of the inherently biological nature of humanity and our co-evolution as integral elements of the complex natural ecosystems of this planet. They chart growing societal and personal unease and/or illness resulting from loss of access to the natural world, and to modern habits and practices that are inconsistent with and which consequently degrade the supportive environments and communities with which we evolved interdependently. The significance of contact with the natural world for recovery from illness has also become increasingly recognized over the past two decades, which has close resonances with the now outmoded concept of convalescent care. For example, in a now famous study in the 1980s,[6] patients recovering from gall bladder removal in a suburban Pennsylvania hospital who had views of nature were found to have statistically significant improved recovery rates relative to those without such views. The phenomenon of health benefit through contact with nature has since become far more generally accepted into the mainstream of healthcare. The 2002 'Wanless report', *Securing Our Future Health: Taking a Long-Term View*,[7] commissioned by the UK Treasury, emphasizes the importance of the environment for health, not merely in terms of factors such as clean air and water but also access to green spaces as well as 'green exercise'. Statistically significant physical and mental health benefits have been shown to accrue from 'green exercise',[8] access to 'green space' is one of the UK government's headline sustainability indicators,[9] and it is no coincidence that recent UK government policy on sustainable development has centred upon 'quality of life'[10] with the emergence of concepts such as 'environmental equity'.[11]

Incrementally, Western science and philosophy are beginning to acknowledge that the dependence between humanity and the ecosystems with which we co-evolved is not only central to our basic biological needs but, as many other cultures around the world accept as axiomatic, vital to our mental and spiritual well-being. All of this, of course, echoes the 'ecosophy' of Norwegian philosopher Arne Naess,[12] the father of 'deep ecology', which encapsulates the understanding that all life has equal intrinsic value, irrespective of the uses to which it may be put by humans, and that the well-being of people and society depends on us living consistently with this reality. We can, indeed, trace this evolution of ecocentric consciousness back through the centuries, as exemplified, for example, by the symbolic poetry of William Blake (1757–1827), and particularly in the evocative and often repeated lines of his 'Auguries of Innocence': 'To see a world in a grain of sand, / And heaven in a wild flower, / Hold infinity in the palm of your hand, / And eternity in an hour'.

Turning the soil

Many aspects of our cultural journey have separated us from a mutually supportive relationship with the land and ecosystems that provide for our needs. Contemporary resource consumption, technologies and population growth are resulting in a wide range of interrelated chemical, physical, ecological and ethical problems resulting from a narrow historic focus on the ways in which we have used land, landscapes and other natural resources as well as technologies to overcome discreetly identified problems ranging from food shortages to control of diseases and predators. Cumulatively, we are eroding the very fabric of the Earth and its capacities to support our physical, mental, social and spiritual well-being. These represent challenges defining the next evolutionary steps for humanity, which must necessarily develop from different world-views to those causing the problems for which we now need solutions.

The relationship between humanity and the land that has shaped the course of civilizations, supported and responded to the pressures of contemporary lifestyles, and reacted to their associated modifications and pressures will be fundamentally important in this future quest for sustainability. The quality of the land and landscapes we inhabit is a primary indicator of our prospects, as our well-being depends very directly upon the health of land, landscapes and broader ecosystems. As Thomas Berry, the cultural historian and ecotheologist, puts it, 'Human health is a subsystem of the Earth's health. You cannot have well humans on a sick planet. And that is what we are trying to do, with all our technologies: we are trying to have well humans on a sick planet.'[13]

These supportive ecosystems have shown remarkable resilience in the face of the decimation of tree cover, fish stocks and many other metrics of biodiversity, acceleration of soil erosion, contamination with all manner of pollutants, and perturbation of water flows and climate. They have progressively adapted to remain the figurative and literal 'fertile soil' in which the seeds of our current and future potential grow. Yet, resilient as they have been, the need for wise stewardship and fair sharing of these irreplaceable resources is pressing if we are not to stretch them beyond breaking point, resulting in progressive or catastrophic breakdown of their integrity and supportive capacities.

We are both rational and spiritual beings. The pan-cultural association that we may feel with the land and its ecosystems defines part of us, and increasingly our scientific rationality is providing robust evidence that lifestyles which unravel the irreplaceable services that the earth provides for us can only lead to mutually assured destruction. Equally, progressive understanding of sustainability pressures and their solutions as well as appropriate and proportionate responses provide us with the means to find new and sustainable ways of living and sharing the Earth with each other. Progress by rational decisions can assure us of moving in the right direction, but deeper ethical forces are vital to compel us so to do. Perhaps both are essential for our salvation?

A new 'land ethic'

In his seminal 1949 work *A Sand County Almanac*,[14] published the year after his death, Aldo Leopold sowed some of the more significant seeds of modern environmental awareness, focusing on concepts of 'right' and 'wrong' and the extension of ethics. Leopold made the observation that ethics are also ecological concepts, stating that 'An ethic, ecologically, is a limitation on freedom of action in the struggle for existence' and could be used to differentiate social from antisocial conduct. This highlighted to Leopold the lack of any ethic '... dealing with man's relation to land and to the animals and plants which grow upon it' and their consequent degradation under the forces of the economy. Leopold recognized the extension of ethics to the environment as essential for our eventual survival and well-being, a sense implicit in much subsequent environmental writing and progress. Leopold's 'land ethic' '... simply enlarges the boundaries of the community to include soils, waters, plants, and animals, or collectively: the land', which is also recognized as implying respect for fellow humans sharing that most fundamental of resources.

Echoes of Leopold's words from more than sixty years ago reverberate today in the consequences of a lack of shared ethos across all sectors of society, including threats inherent in the rapid degradation of supportive ecosystems – land and landscapes, marine fisheries and oceans, the atmosphere and climate, biodiversity, fresh water and aquatic ecosystems, and so forth – and their likely consequences for human misery and conflict. This fundamental conflict between ecological systems and human cultures highlights the need for a central ethical framework around which all can agree and then act concertedly and decisively across policy areas as diverse as agriculture, forestry and other contested uses of land, fisheries, transport, energy, water and the manufacture and disposal of products. This implies not a wholesale abandonment of social and economic progress because we now recognize the beauty and inherent value of nature, but that we need to bring these ecocentric and ethical perspectives into the governance and economic framework by which we will steer our future course.

Ethical frameworks

We need a common, uniting framework upon which all can agree to determine what is 'right' and what is demonstrably 'wrong' with respect to the diverse impacts of land uses, policies, technologies and economic instruments on our shared environment, and their inevitable consequences for all who share and depend upon them. In common with governance systems, global society is not short of belief systems, many of which drive divisions and conflict between their adherents. Therefore, we would be naive to hope to find a single faith or other ideology around which all could rally as a source of ethical guidance. It is therefore essential that we escape the shifting sands of entrenched, ephemeral or divisive cultural, political or religious beliefs.

Establishment of such a uniting framework was implicit within the original definition of sustainable development in the 1987 'Brundtland Report'.[15] Some of the purity of this vision has subsequently become 'lost in translation' under many subsequent redefinitions such that, for those joining the sustainability debate later, the initial clarity has been muddied substantially in the intervening years. Within two years of publication of the 'Brundtland' definition, around 140 alternative and variously modified definitions of 'sustainable development' emerged, and in 2007, 'Currently, it has been estimated that some three hundred definitions of "sustainability" and "sustainable development" exist broadly within the domain of environmental management and the associated disciplines which link with it, either directly or indirectly.'[16] Many of these alternative definitions seem fit only to justify 'business as usual' and protect vested and generally counter-sustainable interests, and arguably may have been created for that purpose. Nevertheless, the basic premise of the three major interdependent themes of sustain-ability – ecological, economic and social systems – remains as pertinent and well supported today, fulfilling part of the need for that uniting ethical position, grounded in testable and consensual principles inclusive of the diversity of landscapes and human cultures that they support. 'Continued degradation of ecosystems is therefore an infringement of human rights at all scales from the local to the global. Sustainable development is thus as much a moral as a biophysical imperative ...'[17]

The axiom of science

Despite long-standing global consensus about the basic tenets of sustainable development, further practical and transparent tools are essential to support the complex and contentious decisions required to turn its aspiration into practical agreements and concerted action. Again, we have to turn to the fundamental laws and principles that govern these essential ecosystems, as revealed by testable and consensual science rather than more culturally relative commodities such as faith or politics, to provide a solid foundation upon which to establish higher ethical standards that may give substance to a collective vision of sustainability.

To some, conflation of science with ethics may sound dangerous, the defensive argument being that science is essentially value free. However, this itself is an outmoded concept in the light of the bias of funding choices and other influences on the kinds of scientific explorations that are and are not conducted,[18] as well as modes of exploration and conclusions that are accepted in the peer-reviewed literature. Conversely, it is far from unreasonable that objective and testable theories and facts should provide a touchstone for what is manifestly 'right' or 'wrong' with respect to the human and ecological consequences of the ways in which we treat and share resources. From such a science-based per-spective, we can explore, debate and reach consensus about how our actions or activities may impair supportive natural systems in ways that compromise their capacities to support any or all dimensions of human well-being. This has very clear moral implications. Ethically, then, if we can form agreement across sectors of society and cultures about what is 'right' and what is 'wrong' in terms of implications for natural carrying capacity as well as opportunities for others to access natural resources,[19] then we will have made significant strides towards practical solutions.

Founded substantially upon the science-based model of sustain-ability of The Natural Step (TNS),[20] which is derived from basic, non-contentious natural laws and principles (such as the laws of thermodynamics, cell biology and basic ecosystem principles), a number of priorities have been proposed to progress towards a better ethical position, precisely because ecosystem and societal breakdown will ensue if they are not addressed.[21] These priorities include overcoming such prevalent 'wrongs' as: an economic system dependent on growth in material consumption within a world of finite resources; physical degradation of biological systems; inequitable use of resources; failure to recognize the basic rights of others; deprivation of access to natural capital by future generations; and denial of credible evidence of environmental change. Each of these priorities touches on one or more principles of sustainability. Identifying each as a wrongful action, based on clear implications for ecological and human consequences, represents a more helpful approach to stimulating changes in the prevailing ethical framework than bald statements about underlying scientific principles and pathways. All of these priorities also help reinforce the view that any economic system must be viewed as a subset of, and unavoidably dependent upon, the integrity of natural systems, rather than the other way round. From this analysis, its authors develop a moral stance rooted in the thermodynamic and ecological realities of the world's underpinning life-support systems, reproduced in Box 10.1.

Real biophysical consequences inevitably stem from denial of opportunity or imposition of inequitable resource access in the absence of clear ethical guidance, with scientifically founded principles of sustainability serving as unshifting foundations for ethical and equitable codes. In addition to TNS, a number of other science-based sustainability systems models ('Five Capitals', STEEP, and so forth) would also undoubtedly prove helpful in framing decisions about the ethical basis of impact upon ecosystems and those, including future generations, who share them. In this context, it is possible to conceive of a common scientific basis for negotiation and agreement, applicable across cultural divides, about the ethics of practices and decisions.

Given the scale of modification required if society is to attain sustainability, the need for supporting tools to stimulate innovation cannot be overestimated. Indeed, to add to our ethical framework, we could state that negative campaigning about environmental 'wrongs' in ways that stifle or fail to encourage innovation of 'rights', or the development of tools that do not promote such innovation, are themselves 'wrongs'.

These ethical dimensions can only amplify in importance as systemic problems, ranging from climate change and burgeoning human population to growing per capita material expectations, impose growing pressures on land and landscapes. This ethical framework highlights,

Box 10.1 Moral stances rooted in thermodynamic and ecological principles

- Abuse of political or economic power, in either a national or international context, is wrong.
- Incautious, permissive environmental discharge regulation is wrong, as it does not steer society towards zero harmful emissions.
- Use of compounds that are relatively persistent and foreign to nature, outside of tightly controlled closed-loop systems, is wrong.
- Use of fossil fuel and other mined materials beyond natural reassimilation rates (which operate over geological time-scales) is wrong.
- Use of nuclear energy, given the accumulation of waste over long timescales and the concomitant costs imposed upon future generations, is wrong. An additional negative factor is the tight linkage of nuclear power to nuclear arms.
- Use of GE (genetically engineered) organisms in uncontained systems is wrong since it threatens to override natural barriers to gene flow.
- Overfishing, forest destruction and over-abstraction are wrong as they irreversibly erode the natural capital upon which the global ecosystem and human well-being depend.
- Failure to investigate the compliance of current practices with System Conditions is wrong, since all participants influence the system as a whole, and sustainability relies on each participant taking responsibility for the system as a whole.

Scource: from Johnston et al. 2007

without any faith-based absolutism, that reducing human environmental impact will inevitably need to attend to population control and distribution, resource consumption patterns, the integrity and 'carrying capacity' of ecosystems, and the distribution of affluence and privilege as key determinants of the ability of human societies as a whole to protect their security and quality of life through progress towards sustainability.

Governance for sustainability

The fact that we share a finite, common pool of natural resources that is dwindling in extent and quality as rapidly as our population and its demands are booming needs to focus our legislation, the execution of the common law, economic appraisal, and pragmatic tools that enable us to make better decisions in an open and democratic manner. However, this legal and supporting infrastructure, familiar though it is to those of us in developed nations, is just one particular model for governing rights to land and other ecosystem resources. In general, it has been applied in a top-down manner in the past. But it is also essential that we move beyond the expectation that legislators and litigators will 'sort it all out' for us.

Perhaps because of the class structure that historically stratified decision-making, UK and European societies as well as the nations that our empires and/or adopted legal systems and other cultural models have shaped, we appear culturally conditioned to abrogate decisions relating to resource conflicts upwards to government or the courts. Indeed, proud though the UK and the USA are to champion democracy to the rest of the world, its internal implementation in each nation is as an executive model wherein elected leaders supported by unelected officials and economically powerful funders make 'expert' judgements on behalf of the wider population. Questions remain as to the breadth of vision of the best interests of the whole of the electorate from this executive model, as well as its impartiality and freedom from influence from vested interests. For example, there were many vociferous critics of the influence of the arms and petrochemical industries on fiscal and foreign policy throughout the ostensibly democratic American presidency of George W. Bush from 2001 to 2009,[22] and of the relationship between political donations, the awarding of civic honours and influence on legislation and policy priorities in the UK over the same period. Since considerable financial support and patronage are required to enable a person to 'climb the greasy pole' towards political office in most, if not all, democratic nations, many observers question how successfully our politicians can represent the best interests and everyday conflicts of diverse electorates that they are empowered and paid to serve. Despite an active culture of protest through open media, particularly in the UK with its vocal and diverse NGO sector, through which various special interests are represented, society seems less ready to participate at a personal level in interactive governance processes. Though manifestly democratic, our established executive model of government is far removed from a participative model of governance in which all people feel involved and which leads to decisions towards which they feel they have contributed within an equitable framework. This is one of the aspirations of the UK's 'Big Society' experiment, instigated as a core policy driver in 2010 (see the previous chapter).

A further critique of the experience of executive democracy is the extent to which the people that this executive is intended to serve are involved in the process of voting, whether by direct disenfranchisement or apathy engendered by low expectation of effecting change. As an extreme, apartheid is a striking example of a democratic government regime based on the views of only that stratum of society deemed worthy of a vote. The second and third decades of Robert Mugabe's regime in Zimbabwe were also ostensibly democratic, albeit fraught with the dangers inherent in the absence of an effective opposition and the later exclusion from political influence of the Movement for Democratic Change (MDC) party as it grew into a stronger opposing force. In reality, of course, Zimbabwe's democratic veneer covered a tribal leadership model in which the 'chief' centralized power and grew increasingly resistant to challenge, conflicting world-views, domestic economic and epidemiological crises, and eventually the disapprobation of the international community. This same tendency to centralization of power has also been widely seen in communist regimes, as indeed in parliamentary or presidential democracies, which appeared with hindsight merely to entrench the interests of an oligarchy rather than to respond to the guidance of a representative meritocracy.

However, if the term 'democracy' is instead taken to include true representation of the interests of all stakeholders in a society that is equitable and sustainable in its policies and decisions, there is an overriding need to reflect the views and needs of all stakeholders in decisions pertaining to the far-sighted stewardship and sharing of the natural resources essential to their continuing well-being and opportunity. No country has yet achieved this, despite our many 'experiments' with participative, representative and executive forms of democratic systems across the world. Perhaps the best example today of a commitment to these truly democratic ideals is in post-apartheid South Africa, as exemplified (and described variously elsewhere in this book) by the National Water Act 1998 with its clear intent for water allocation to become explicitly instrumental in the country's overall economic development goals in which poverty eradication and redress of historic inequities figure prominently.

Human societies throughout time and space have experimented, by design but more often by happenchance, with a great diversity of other models of governance. World cultures have operated on the basis of democracy in a bewildering variety of flavours – matriarchal and patriarchal tribes, autocracy and oligarchy, communism, councils of elders, anarchy, and theocracy and royalty, both with absolute or titular figureheads. Although world religions espouse higher ideals, individual advantage and the disempowerment of large swathes of societies assumed to be supplicant to religious authority – particularly women, minorities and other vulnerable constituencies – are common features of many theocracies from the Mogul Empire to the Holy Roman Empire of medieval Europe and the modern-day Taliban. The irony is that, although models such as communism and certain religious faiths are hypothetically opposed to hegemony, the centralization of power and resources by a privileged governing cadre of society has often been the net outcome for many such governments. Throughout history, theological and atheist ideals alike have tended to entrench advantage and power in ruling elites, disempowering large swathes of the societies that they profess to serve.

A governance model that is today learning from its history to become effective in implementing the ecosystems approach is communist China. In recent years, the Chinese government has recognized that the massive flooding of the Yangtze river in 1998 that displaced 120 million people was not, as initially described, an 'act of God' (which itself is an interesting metaphor in a purportedly secular society). Instead, massive deforestation of the upper Yangtze catchment was found to be a major contributory factor to both the amplification of flood peaks and the loss of productive soils through erosion.[23] Furthermore, peaky flows and high sediment loads in the Yangtze are now perceived as prejudicing navigation, hydroelectric generation and the longevity of the Three Gorges Dam and other impoundments lower down the river system. Since one tenth of the world's human population lives in the Yangtze basin, protection or restoration of the ecosystem services that support the needs of this mass of humanity is critical. As part of China's response to this challenge, a massive reforestation programme is under way in the upper catchment of the Yangtze, some of which falls under the Natural Capital Project: China Demonstration Site,[24] which will also sequester a substantial mass of carbon. The upper Yangtze today hosts the world's largest ecosystem service-based land-use change programmes and investment in science and modelling in the shape of the National Forest Conservation Programme (NFCP, referred to as the 'Natural Forest Conservation Programme' in some sources) and the Grain To Green Programme (GTGP), instigated in 2000, both of which aim to halt soil erosion and to restore the hydrology of severely modified upper catchments and so constrain the huge social and environmental costs that have resulted downstream. The political regime in China is both authoritarian and attuned to making decisions in top-down ways judged to be in the wider public interest, with all land owned by the state. However, this does streamline the path towards a rapid and extensive mobilization of Chinese land use in the face of clear and mounting ecosystem-derived problems. Agreements on use of the land are allocated to local people by the state, and these people are only now beginning to engage in markets for land-use rights. Therefore, the current lack of a strong culture of private landownership and rights may facilitate this top-down mandating of landscape-scale restoration, though it is inevitable that the livelihoods of many in the upper Yangtze catchment will be profoundly altered by such massive interventions. Information on the means by which these local people are engaged in land-use change decisions as interested stakeholders is elusive. Given the depth of established property rights in much of the rest of the industrialized world, and a strengthening tradition of stakeholder engagement in policy decision-making and implementation, it is difficult to see more widespread uptake of the Chinese model of enforced transition in land management for ecosystem services of greatest public benefit, albeit that the ideal of ecosystem-based solutions is an important cultural aspiration.

Beyond these centralized government models, there is a diversity of common stewardship approaches to the management and sharing of natural resources.[25] In rural India, land owned by distant interests or else of contested ownership is often shared as a common property by village communities, which collaborate in the production of crops and stock, with protected river reaches and habitat within sight of temples effectively serving as common nature conservation areas. A wide variety of communal stewardship models is also found across the world, ranging from international conventions to a diversity of tribal governance models, down to local level, such as commoners' rights with respect to grazing and hay cropping on commonly held land throughout the UK, such as the Lugg Meadows (Herefordshire) and North Meadow (Cricklade, Wiltshire), both of which are of high nature conservation value and are managed around dates set by the Church calendar but adapted to year-on-year climatic and growing conditions through the decisions of councils of commoners. Examples of international conventions include agreements on exploitation of international fisheries, the more broad-ranging United Nations Convention on the Law of the Sea, through to regional multinational agreements such as the EU's Common Fisheries Policy.

Another governance model, tribal stewardship, has many commonalities across the world. For example, traditional Zulu tribal lands in South Africa are governed through a council of elders (*izindunas*) under the overview of a chief or king. More rarely, a king may rule directly, such as in modern-day Swaziland. In all cases, elders, leaders or kings arbitrate on decisions according to established social codes, with customary protocols for the sharing of critical resources such as productive land and water. South Africa demonstrates starkly contrasting examples not only of patterns of tribal stewardship but also their coexistence with the dominant developed-world private landowning model. Other tribes are nomadic, such as the Maasai of Kenya and Tanzania, with no deeply rooted concept of landownership but rather migrating as a pastoral community with the availability of grazing.

It is difficult to see these models of landlessness or common property forming a viable basis for governance of broader or national public interests in land and its functions in more developed countries, beyond the localized, historically rooted and generally small 'commons' that escaped the enclosure system and have endured ever since. Neither is it apparent that the utopian ideal advanced by some commentators of abandonment of private property rights as a pathway to sustainable development through reversing the concentration of wealth and power in a minority[26] is either workable or realistic.

Superimposed over the variants of democracy and other governance models across much of the modern world is the doctrine of capitalism, by far the world's most pervasive cultural paradigm and certainly one with greater and deeper reach than any religious or other fundamentalist ideal. In order to make sustainable progress, short of a revolution and acknowledging the deep entrenchment of landowning culture, it is pragmatic to assume that we will need to work within and progressively develop such dominant systems, which shape our world. Under capitalist ideology, 'the market' is today imbued with an almost numinous quality, in which financial progress and profit have often been construed as synonymous with societal well-being, or at least the ability to purchase it. There is, of course, an inherent contradiction in this assumption, as the economy ultimately rests upon trading of tangible 'goods and services', production of which by supporting ecosystems is bound by hard biophysical limits. Furthermore, the market is inherently amoral and, as such, if unconstrained, will tend to liquidate natural and human capital to maximize financial return in the short term. This is not to say that the market is automatically a 'bad thing'. If the market in its reductively blinkered form has the power to create the scale of problems we face today then it also has the potential to metamorphose into as powerful an agent of sustainable change if imbued with ecological and ethical principles to set the creation of wealth within a broader context of sustainability and the optimization of equitable and enduring human value. Many authors have discussed how to act on this conclusion from the perspectives of corporate interests,[27] engaged consumers[28] or markets connecting both.[29] When it comes to valuing and internalizing the core ecological and social capital that underpins economic activities, we really have not yet tried out the full promise of capitalism.[30]

The reason for this amble through global governance systems is to reveal that, beneath all this cultural complexity and diversity, it would be facile to hold up any established political model as a panacea for equitable stewardship of land, landscapes and other natural resources underpinning societal well-being and a more sustainable world. However, what does emerge from this overview of alternative governance systems is the central importance of a strong social infrastructure around which to create consensus about the sharing of, and caring for, common natural resources. This conclusion was also formed by Jared Diamond in considering a wide range of failing or successful civilizations throughout human history and across the world in his alarming and informative book *Collapse*.[31]

The new democracy

The scope and nature of sustainability issues facing the world today are often framed as environmental problems. In fact, they are in no way environmental problems; the environment is often just the medium by which societal pressures and decisions are unavoidably circulated back to impinge on the health and potential of people. 'Environmental' issues and the policies implemented to address them start and end with 'people', whether as polluters or polluted, exploiters or victims of degraded ecosystems, or as stewards of environmental quality. The environment will adapt and continue in one form or another, degraded or not; the big question is really whether or not we choose and act to continue to live fulfilling lives in that future, supported by ecosystems of sufficient scale, diversity and resilience.

Sustainable development is a challenge that depends upon the involvement, common understanding and collective innovation of all sectors of society, not just the top-down strata of society to whom we may have looked for leadership in times gone by and which we may wrongly assume to have both the necessary power and the answers to our problems. All of us in society shape our future daily through our behaviours. From local to global scales, it is people that are at the root of both problems and their solutions. Each and every one of us has a stake in the fate of the common resources upon which we will depend, and therefore in how we act now to safeguard them. Each of us has the opportunity to effect change in how we think and act, every day-to-day opportunity forming another 'bifurcation point' that can propel us towards or away from sustainability. Our values are manifest in our actions rather than words alone, defining the respect with which we hold ourselves, our neighbours and our collective future.

The central role of people in decision-making, and the need to account for the livelihood needs of all sectors of society, are basic requirements, albeit ones that do not automatically play well with many top-down governance systems found across the world. Equally, the complete devolution of judgement to the local scale can result in decisions that fail to respect the integrity and functioning of high-level ecosystems. Fully sustainable governance therefore has to account for both the needs of local people and the broader well-being of whole landscape ecosystems, catchments, islands, broad-scale ecosystem processes, continents, and the integrity of the global atmosphere – and indeed the whole biosphere. Involving people such that they feel a sense of ownership in far-seeing decisions reflecting both their diversity and long-term interests is a far from simple matter. The shift in recognition and safeguarding of public as well as private benefits accruing from the use of land, reviewed extensively in Chapter 2, is a crucial step towards inclusive governance. There is a need better to inform governance decisions with both the perspectives of science and other expert input but also the needs and diverse value systems of all people in society, not just those represented by the perspectives of a governing cadre. Many commentators have outlined the need for a 'civic science' model in which scientific and other 'expert' views are linked with other forms of knowledge held by society to make better-informed decisions,[32] and which Anthony Turton elegantly sums up as a 'trialogue' of science, governance and society.[33]

This brings us back to Colbourne's alternative, stakeholder-based EDD (engage-deliberate-decide as outlined in Chapter 9) model of decision-making, with decisions resulting from patient processes of engaging all relevant stakeholders in dialogue, leading to collective deliberation and the co-creation and evaluation of options from which final decisions may emerge. The up-front costs and time requirement of the EDD process are substantially greater than merely allowing experts to make decisions. However, once a decision has been reached through the EDD process, it will tend to be more innovative, adapted to multiple local needs and stakeholder perspectives, and so accepted by people who feel that they have a stake in the outcome, even if it does not completely satisfy all of their requirements. By contrast, if we look at the increasingly massive defensive activities relating to expert-derived proposals, for example in major road schemes, additions of runways to airports, port enlargement or similar major development schemes that the public may feel are imposed upon them, it then becomes apparent that defensive post-decision life-cycle costs and timescales resulting from the DAD model will tend vastly to exceed up-front investment in the EDD process.

There are, of course, variants, exceptions and extensions to this EDD model, including circumstances in which the DAD (decide-announce-defend) model is appropriate. For example, where quick decisions are needed with respect to matters that are non-contentious or for which basic principles are well understood or tested (for example, in sending an ambulance to a road accident or in permitting a land use for which a higher-level policy has been agreed consensually) then DAD serves that purpose efficiently. Also, increasing levels of consultation can be applied to the authoritarian DAD model to help mitigate impacts and ease acceptance by society and, although this 'DAD-lite' model still depends largely on initial expert framing of decisions with only *post hoc* stakeholder moderation, it may nevertheless provide a quick and cheap means to support decisions that are not deeply contentious, long-term or affecting large numbers of people. A practical example of this that we have already considered is spatial planning decisions where informed by prior policy agreements and transparent tools such as the insightful *eThekwini Catchments: A Strategic Tool for Management*.[34] Conversely, in complex and potentially contentious situations, the full EDD process may be required, noting that this may also set precedents which can, with care, form future consensual protocols to guide expert (DAD or 'DAD-plus') decision-making better reflective of wider stakeholder values. This, in some ways, mirrors the development of case law under the common law, which has a history of responding relatively quickly to evolving societal values and expectations, and which can subsequently inform the continued development of statutory regulation. The guiding principle here is that investment in the decision-making process needs to be proportionate to the risks inherent in the decision.

Top-down legislation serves an important purpose as an overall framework to guide practical decision-making and, like South Africa's visionary water laws, which have been used as an exemplar at various points in this book, can also mandate EDD-style stakeholder engagement processes to help support devolved, localized decisions and also shape new institutions relating to the allocation of water as a practical and powerful means to help the nation move from its history of apartheid segregation and privilege towards an inclusive and democratic society. It is, in practice, impossible to conceive of equitable and sustainable decision-making relating to land, landscapes and other critical natural resources without some form of active stakeholder engagement. This is governance for the people, by the people, supported but not dominated by wise government and its experts, and respecting the critical natural capital that supports the needs of all.

The radically different, deeply interconnected world-view offered by ecosystem services enables us to further open up the debate about what exactly constitutes appropriate sustainable lifestyles, taking account of the needs of all stakeholders, including future generations, and the functional, beautiful and cherished ecosystems that they require to allow them to fulfil their potential.[35] It thus opens the door to a pure form of democracy that respects the rights and potential of all of humanity and the well-being of the Earth upon which it depends. Ecosystem services, then, are not only a powerful practical tool but also a potentially subversive instrument, challenging world-views, world orders, everyday assumptions and vested interests established throughout more than two centuries of industrialization and resource exploitation largely blinkered to their associated and largely unintended 'collateral damage'. They provide us with a science-based yet practical and readily understood means to reflect the ethics of ecosystem protection in ways that relate to both social and economic security and progress.

If democracy is ultimately about 'one person, one vote' then it is implicitly about not only having a voice but also listening to all other voices. It is about the part we must play in shaping a just and sustainable future, which ultimately rests upon how we share fundamental ecosystem resources, ranging in scale from local habitat to landscapes, water and catchments, regions, nations and the whole biosphere. It is our land, our heritage. We win together or we suffer together; the choice is that simple as we struggle to find a workable democratic model.

Down to Earth

Land and landscapes are far from remote academic concepts, merely bearing our weight and growing our crops and to be cared for in some vague altruistic way. They are instead what we eat, the places where the food is grown and which yield fresh water and purify air, which maintain our physical structure, and which constitute the matter that forms our homes and computers. The wider environment is no diffuse thing but is what we breathe deeply into our lungs and which percolates into our bloodstreams to nourish all the cells of our body, and is the water that we drink, cleansing, supporting, dissolving and conveying the substances that make us what we are. It is something irreplaceable that we depend upon in our use of energy, our recreational, spiritual, educational and cultural pursuits, and our enjoyment and expectation of living fulfilled and useful lives. The environment's inherent properties regulate the risk of our homes being flooded and provide the raw materials ultimately underpinning all of our economic activities from primary resources through to waste assimilation. The ripples of our actions and decisions, all of them inescapably affecting the ecosystems of which we are indivisible parts and which support our needs and aspirations, inevitably roll out to influence the interests of all who share this single world. One Earth, one people, one interdependent whole; one chance to realize our collective future.

What we need now is the humility, wisdom and courage to act upon what we know to forge a different world, if not through altruism or grand ideals then in recognition that equity and sustainability are the basic prerequisites for a decent long-term future for ourselves and those we love. Our land and landscapes and the ecosystems that define them are both our inheritance and our legacy, defining what is possible for humanity and also informing us about ways of living that offer us all unrestricted potential. From scientific, philosophical, spiritual, psychological and basic biological realities, there springs a clear moral imperative to cherish, manage and share, fairly and wisely, this 'common ground' that unites us all through space and time.

Notes

Acknowledgements

1 J. Colvin, F. Ballim, S. Chimbuya, M. Everard, J. Goss, G. Klarenberg, S. Ndlovu, D. Ncala and D. Weston (2008), 'Building capacity for co-operative governance as a basis for integrated water resources managing in the Inkomati and Mvoti catchments, South Africa', *Water SA*, 34(6): 681–90.

Introduction

1 C. Troll (1939), *Luftbildplan und ökologische Bodenforschung* [Aerial photography and ecological studies of the earth], Zeitschrift der Gesellschaft für Erdkunde, Berlin, pp. 241–98.

2 M. G. Turner (1989), 'Landscape ecology: the effect of pattern on process', *Annual Review of Ecology and Systematics*, 20: 171–97.

3 M. Allaby (1998), *Oxford Dictionary of Ecology*, Oxford University Press, Oxford.

4 J. Sanderson and L. D. Harris (eds) (2000), *Landscape Ecology: A Top-Down Approach*, Lewis Publishers, Boca Raton, FL.

1 The privatization of the land

1 G. Cowan (2002), *Nomadology in Architecture: Ephemerality, Movement and Collaboration*, University of Adelaide, hdl.handle.net/2440/37830 [1]. Nomadic communities today break down into the two main categories of pastoralists and non-pastoralists, the latter of which are often the most neglected in government systems. Many non-pastoralists pursue hunter-gathering, the oldest human practice in the world. For example, Taureg people are masters of the inhospitable Saharan landscape, eking out an existence with little contact with the economically developed world. Pastoralists, such as many among the Maasai tribe of East Africa, who live migratory, low-impact lifestyles responding to the availability of good grazing, often confront more 'developed' societies only when they find themselves excluded from formerly uninhibited access to water and grazing. Land claims and barbed wire are indeed the frequent hallmarks of the acquisitive tendencies of the economically developed world, which generally fails to recognize other value systems and livelihoods such as those practised through millennia by nomadic people. A minority of the world's nomads are 'peripatetic nomads' with lifestyles based on trading, moving between markets and customers, and so are more closely tied to the margins of the industrialized world.

2 E. Moussavi-Nejad (2003), 'Censuses of pastoral nomads and some general remarks about the census of nomadic tribes of Iran in 1998', *Nomadic Peoples*, 7(2): 24–35.

3 This estimate was produced by the 'Renke Commission', the National Commission for Denotified, Nomadic and Semi-Nomadic Tribes (NCDNSNT), set up under India's Ministry of Social Justice and Empowerment; www.ncdnsnt.gov.in.

4 A commonly cited version of Chief Seattle's speech was published in the *Seattle Sunday Star* on 29 October 1887 in a column by Dr Henry A. Smith. However, there is controversy surrounding Chief Seattle's speech, with various other published versions and, indeed, debate over its very existence.

5 This definition is adapted from that used by Natural England, the statutory nature conservation regulator in England; www.naturalengland.org.uk.

6 This definition is adapted from the Department for Environment, Food and Rural Affairs (Defra), the English government department with responsibility for the environment, including its agricultural uses; www.defra.gov.uk.

7 J. Diamond (1998), *Guns, Germs and Steel*, Viking Penguin, London.

8 S. J. B. Cox (1985), 'No tragedy on the commons', *Environmental Ethics*, 7: 49–61; C. J. Dahlman (2008), *The Open Field System and Beyond: A Property Rights Analysis of an Economic Institution*, Cambridge University Press, Cambridge; T. Dietz, E. Ostrom and P. C. Stern (2003), 'The struggle to govern the commons', *Science*, 302: 1907–12.

9 G. Hardin (1968), 'The tragedy of the commons', *Science*, 162: 1243–8.

10 A. J. L. Winchester (2008), 'Statute and local custom: village byelaws and the governance of common land in medieval and early-modern England', Paper for the IASC Global Conference, Cheltenham, July, http://dlcvm.dlib.indiana.edu/archive/00004046/.

11 J. Bettey (1999), 'The development of water meadows in the southern counties', in H. Cook and T. Williamson (eds) (1999), *Water Management in the English Landscape: Field, marsh and meadow*, Edinburgh University Press, Edinburgh, pp. 179–95.

12 C. P. Rodgers, E. A. Straughton, A. J. L. Winchester and M. Pieraccini (2010), *Contested Common Land; environmental governance past and present*, Earthscan, London.

13 E. Ostrom (1990), *Governing the Commons: The evolution of institutions for collective action*, Cambridge University Press, Cambridge.

14 G. Slater (1907) *The English Peasantry and the Enclosure of Common Fields*, Archibald Constable and Co. Ltd, London; J. L. Hammond and B. Hammond (1911), *The Village Labourer, 1760–1832: A Study in the Government of England before the Reform Bill*, Longmans, Green and Co., London.

15 J. L. Hammond and B. Hammond (1917), *The Town Labourer*, Longmans, Green and Co., London.

16 E. P. Thompson (1991), *The Making of the English Working Class*, Penguin, London.

17 D. Brockington (2009), *Celebrity and the Environment: Fame, wealth and power in conservation*, Zed Books, London and New York.

18 J. R. Wordie (1983), 'The chronology of English enclosure 1500–1914', *Economic History Review Series 2*, 36(4): 483–505 (quote from p. 489).

19 W. A. Armstrong (1981), 'The influence of demographic factors on the position of the agricultural labourer in England and Wales, c 1750–1914', *Agricultural History Review*, 29: 71–82; J. D. Chambers and G. E. Mingay, *The Agricultural Revolution 1750–1850*, Batsford, London, reprinted 1982.

20 F. Engels (1882), *Die Entwicklung des Sozialismus von der*

Utopie zur Wissenschaft, Hottingen, Zurich.

21 J. Cock and E. Koch (1991), *Going Green: People, Politics, and the Environment in South Africa*, Oxford University Press, Cape Town.

22 C. Merchant (2007), *American Environmental History: An Introduction*, Columbia University Press, New York.

23 M. Rangarajan (2001), *India's Wildlife History*, Permanent Black, Delhi.

24 K. B. Ghimire and M. P. Pimbert (1997), *Social Change and Conservation*, Earthscan, London; J. Igoe and B. Croucher (2007), 'Poverty alleviation meets the spectacle of nature: does reality matter?', *Conservation and Society*, 5(4): 534–61.

25 Brockington, *Celebrity and the Environment*.

26 UNESCO, www.unesco.kz/heritagenet/kz/hn-english/history_en.htm.

27 D. Zeidan (1995), *The Middle East Handbook: Information on States and People Groups in the Middle East*, vol. 3: *A–Z ME People Groups and their Distribution*, OM-IRC, www.angelfire.com/az/rescon/MEHBKNMD.html.

28 Merchant, *American Environmental History*.

29 M. D. Spence (1999), *Dispossessing the Wilderness: Indian removal and the making of national parks*, Oxford University Press, Oxford.

30 Brockington, *Celebrity and the Environment*.

31 A. Balmford and T. Whitten (2003), 'Who should pay for tropical conservation, and how could these costs be met?', *Oryx*, 37(2): 238–50.

32 World Commission on Dams (2000), *Dams and Development: A New Framework for Better Decision-making*, Earthscan, London.

2 Reclaiming the common good

1 M. Everard and T. Appleby (2009), 'Safeguarding the societal value of land', *Environmental Law and Management*, 21: 16–23.

2 www.un.org/en/documents/udhr/index.shtml.

3 D. A. Posey (ed.) (2000), *Cultural and Spiritual Values of Biodiversity*, ITDG Publishing, Rugby.

4 V. Strang (1997), *Uncommon Ground: Landscape, Values and the Environment (Explorations in Anthropology)*, Berg, Oxford.

5 E. Ostrom and N. Dolšak (2003), *The Commons in the New Millennium: Challenges and Adaptations*, MIT Press, Cambridge, MA.

6 B. van Koppen, M. Giordano and J. Butterworth (2008), *Community-Based Water Law and Water Resource Management Reform in Developing Countries*, Comprehensive Assessment of Water Management in Agriculture Series 5, CABI, Wallingford, www.cabi.org.

7 Republic of South Africa (1998), *National Water Act*, Act no. 36 of 1998, Pretoria, South Africa.

8 I. R. Bowler and B. W. Ilbery (1999), 'Agricultural land-use and landscape change under the post-productivist transition – examples from the United Kingdom', in J. Baudry, I. R. Bowler, R. Kronert and A. Reenberg (eds), *Land-use Changes and Their Environmental Impact in Rural Areas in Europe (Man & the Biosphere)*, Taylor and Francis.

9 R. D. Gregory, D. G. Noble and J. Custance (2004), 'The state of play of farmland birds population trends and conservation status of lowland farmland birds in the United Kingdom', *Ibis*, 146: 1–13.

10 English Nature (2001), *Accentuate the Positive – funding positive management on SSSIs*,

Press release, 2 March 2001, www.english-nature.org.uk/news/story.asp?ID=259.

11 Environmental Audit Committee, Eleventh Report of Session 2005–06, *Outflanked: The World Trade Organization, International Trade and Sustainable Development*, HC 1455.

12 C. Merchant (2007), *American Environmental History: An Introduction*, Columbia University Press, New York.

13 R. A. Posner (2009), *A Failure of Capitalism: The Crisis of '08 and the Descent into Depression*, Harvard University Press, Boston, MA.

14 Time Magazine (1932), 'The Roosevelt week', *Time*, 11 July 1932.

15 K. Asmal (2008), 'Appendix: Reflections on the birth of the National Water Act, 1998', *Water SA*, 34(6): 662–4.

16 P. Hawken (1993), *The Ecology of Commerce: A Declaration of Sustainability*, HarperCollins, New York.

17 J. Porritt (2005), *Capitalism as if the World Matters*, Earthscan, London; J. Bishop, S. Kapila, F. Hicks, P. Mitchell and F. Vorhies (2008), *Building Biodiversity Business*, Shell International Limited and the International Union for Conservation of Nature, London, UK, and Gland, Switzerland, cmsdata.iucn.org/downloads/bishop_et_al_2008.pdf; M. Everard (2009), *The Business of Biodiversity*, WIT Press, Ashurst; TEEB (2010), *TEEB – The Economics of Ecosystems and Biodiversity: Report for Business – Executive Summary*, www.teebweb.org/LinkClick.aspx?fileticket=ubcryEoOUbw%3d&tabid=1021&language=en-US.

18 M. Everard and T. Appleby (2008), 'Ecosystem services and the common law: evaluating the full scale of damages', *Environmental Law and Management*, 20: 325–39.

19 G. Hardin (1968), 'The tragedy of the commons', *Science*, 162: 1243–8.

20 Reviewed, for example, by Everard, *The Business of Biodiversity*.

21 W. Odum (1982) 'Environmental degradation and the tyranny of small decisions', *BioScience*, 32(9): 728–9.

22 E. Ostrom (1990), *Governing the Commons: The evolution of institutions for collective action*, Cambridge University Press, Cambridge.

23 For example, '… an intention or policy of including people who might otherwise be excluded or marginalized, such as the handicapped, learning-disabled, or racial and sexual minorities …' (*Oxford Pocket Dictionary of Current English*, 2009).

24 A. Stirling, M. Leach, L. Mehta, I. Scoones, A. Smith, S. Stagl and J. Thompson (2007), *Empowering Designs: Towards more progressive appraisal of sustainability*, STEPS Working Paper 3, STEPS Centre, University of Brighton; A. Stirling (2008), '"Opening up" and "closing down": power, participation, and pluralism in the social appraisal of technology', *Science, Technology and Human Values*, 33(2): 262–94.

25 B. Bruns (2005), *Community-Based Principles for Negotiating Water Rights: Some conjectures on assumptions and priorities*, International workshop on 'African water laws: plural legislative frameworks for rural water management in Africa', 26–28 January, Johannesburg, South Africa; M. S. Hrezo and W. E. Hrezo (1985), 'From antagonistic to cooperative federalism in water resources development: a model for reconciling federal, state and local programs, policies and planning', *American Journal of Economics and Sociology*, 44(2), April.

26 J. Colvin, F. Ballim, S. Chimbuya, M. Everard, J. Goss, G. Klarenberg, S. Ndlovu, D. Ncala and D. Weston (2008), 'Building capacity for co-operative governance as a basis for integrated water resources managing in the Inkomati and Mvoti catchments, South Africa', *Water SA*, 34(6): 681–90.

27 Millennium Ecosystem Assessment (2005), *Ecosystems and Human Well-Being: Synthesis*, Island Press, Washington, DC.

28 J. Sheail (2002), *An Environmental History of Twentieth-century Britain*, Palgrave Macmillan, Basingstoke; Everard, *The Business of Biodiversity*.

29 W. Devall and G. Sessions (2001), *Deep Ecology: Living as if Nature Mattered*, Gibbs Smith, Layton, UT.

30 R. M. Baum (2010) 'Perspective: sustainable growth is an oxymoron', *Government and Policy*, 88(45): 44–47, DOI:10.1021/CEN110210164212.

31 J. H. Spangenberg and J. Settele (2010), 'Precisely incorrect? Monetising the value of ecosystem services', *Ecol Complex*, 7(3): 327–37.

32 A. Giddens (2006), *Sociology*, Polity Press, Cambridge.

33 WCED (1987), *Our Common Future*, Oxford University Press, Oxford.

34 P. Johnston, M. Everard, D. Santillo and K.-H. Robèrt (2007), 'Reclaiming the definition of sustainability', *Environmental Science and Pollution Research*, 14(1): 60–66.

35 *The UNECE Aarhus Convention: Convention on Access to Information, Public Participation in Decision-making and Access to Justice in Environmental Matters*, www.unece. org/env/pp/.

36 S. R. Arnstein (1969), 'A ladder of citizen participation', *Journal of the American Planning Association*, 35(4): 216–24.

37 L. Colbourne (2009), *Mainstreaming Collaboration with Communities and Stakeholders for FCERM. Improving Institutional and Social Responses to Flooding*, Joint Defra/ Environment Agency Flood and Coastal Erosion Risk Management R&D Programme, Science Report: SC060019, Work Package 4, Environment Agency, Bristol, publications. environment-agency.gov.uk/pdf/ SCHO0509BQBR-E-E.pdf.

38 Green infrastructure in the NW Region, www.gos.gov.uk/gonw/ EnvironmentRural/Environmental Issues/753502/.

39 B. Woods et al. (2007), *The SUDS Manual*, CIRIA Report C697, Construction Industry Research and Information Association, London.

40 M. Everard and H. L. Moggridge (in press), 'Rediscovering the value of urban rivers', *Urban Ecosystems*.

41 J. Benington (2010), 'From private choice to public value?', in J. Benington and M. H. Moore (eds), *Public Value: Theory and Practice*, Palgrave Macmillan, Basingstoke.

3 The ends of the Earth

1 K. S. Zimmerer (1994), 'Human geography and the "New Ecology": the prospect of promise and integration', *Annals of the Association of American Geography*, 84: 108–25.

2 Millennium Ecosystem Assessment (2004), *Millennium Ecosystem Assessment*, www.maweb.org; Millennium Ecosystem Assessment (2005), *Ecosystems and Human Well-Being: Synthesis*, Island Press, Washington, DC.

3 WWF (2004), *Living Planet Report 2004*, WWF, Godalming, UK,

www.panda.org/downloads/general/lpr2004.pdf.

4 UNDP (United Nations Development Programme) (2004), *Human Development Report 2004: Cultural Liberty in Today's Diverse World*, UNDP, New York,. hdr.undp.org/reports/global/2004/.

5 IPCC (Intergovernmental Panel on Climate Change), www.ipcc.ch.

6 J. Diamond (2005), *Collapse: How Societies Choose to Fail or Succeed*, Viking Penguin, London.

7 P. Johnston, M. Everard, D. Santillo and K.-H. Robèrt (2007), 'Reclaiming the definition of sustainability', *Environmental Science and Pollution Research*, 14(1): 60–66.

4 Shifting conceptual landscapes

1 Millennium Ecosystem Assessment (2005), *Ecosystems and Human Well-Being: Synthesis*, Island Press, Washington, DC.

2 T. S. Kuhn (1962), *The Structure of Scientific Revolutions*, University of Chicago Press, Chicago, IL.

3 R. H. Khatibi (2002), 'Systemic knowledge management in hydraulic systems – Paper 1: a postulate on paradigm shifts as a methodological tool', *Journal of Hydroinformatics*, 5(2): 127–40.

4 P. Denny (1994), 'Biodiversity and wetlands', *Wetlands Ecology and Management*, 3: 55–61.

5 M. Kottelat and J. Freyhof (2007), *Handbook of European Freshwater Fishes*.

6 Millennium Ecosystem Assessment (2005), *Ecosystems and Human Well-Being*.

7 P. B. Medawar and J. S. Medawar (1977), *The Life Science*, Wildwood House, London.

8 W. Howarth (2002), 'Editorial: Implementing flood protection policy through planning', *Water Law*, 13: 247–8.

9 P. Ehrlich and A. Ehrlich (1982), *Extinction: The Causes and Consequences of the Disappearance of Species*, Gollancz, London.

10 W. Howarth (2001), 'Water, water everywhere', *Water Law*, 12: 1–3.

11 I. Burton, R. W. Kates and G. F. White (1978), *The Environment as Hazard*, Oxford University Press, New York.

12 J. L. Scrase and W. R. Sheate (2005), 'Re-framing flood control in England and Wales', *Environmental Values*, 14: 113–37.

13 M. Everard, M. Bramley, K. Tatem, T. Appleby and W. Watts (2009), 'Flood management: from defence to sustainability', *Environmental Liability*, 2: 35–49.

14 Scrase and Sheate, 'Re-framing flood control'; RSPB (Royal Society for the Protection of Birds) (1970), *Farming and Wildlife: A Study in Compromise*, Proceedings of a conference held at Silsoe, 1969, RSPB, Sandy, Bedfordshire.

15 M. Everard (2005), *Water Meadows: Living treasures in the English landscape*, Forrest Text, Ceredigion.

16 C. Johnson, S. Tunstall and E. Penning-Rowsell (2004), *Crises as Catalysts for Adaptation: Human response to major floods*, Flood Hazard Research Centre, Publication no. 511.

17 Scrase and Sheate, 'Re-framing flood control'.

18 G. P. Marsh (1864), *Man and Nature; or, physical geography as modified by human action*.

19 M. Everard and T. Appleby (2009), 'Safeguarding the societal value of land', *Environmental Law and Management*, 21: 16–23.

20 L. von Bertalanffy (1940), 'The organization considered as a physical system', Reprinted in *General Systems Theory: Foundations, development, applications*, Braziller, New York, 1968, pp. 120–38.

21 H. Plunket Greene (1924), *Where the Bright Waters Meet*, Republished 2007, Medlar Press, Ellesmere.

22 A. Leopold (1949), *A Sand County Almanac: And Essays on Conservation from Round River*, Oxford University Press, New York.

23 R. Carson (1962), *Silent Spring*, Hamish Hamilton, London.

24 www.audubon.org.

25 Waverley Committee (1954), *Recommendations of the Waverley Committee. Report of the Departmental Committee on Coastal Flooding* (Cmd 9165), HMSO, London.

26 World Commission on Dams (2000), *Dams and Development: A New Framework for Better Decision-making*, Earthscan, London.

27 M. Everard and A. Powell (2002), 'Rivers as living systems', *Aquatic Conservation*, 12: 329–37.

28 J. Cock and E. Koch (1991), *Going Green: People, Politics, and the Environment in South Africa*, Oxford University Press, Cape Town.

29 W. Beinart and P. Coates (1995), *Environment and History: The Taming of Nature in the USA and South Africa*, Routledge, London; J. Mittelman (1998), 'Globalization and environmental resistance politics', *Third World Quarterly*, 19(5): 847–72.

30 P. Kameri-Mbote and P. Cullet (1996), *Environmental Justice and Sustainable Development: Integrating local communities in environmental management*, IELRC Working Paper 1996 – 1, International Environmental Law Research Centre, Geneva, www.ielrc.org/content/w9601.pdf.

31 C. C. Chi (2001), 'Capitalist expansion and indigenous land rights: emerging environmental justice issues in Taiwan', *Asia Pacific Journal of Anthropology*, 2(2): 135–53.

32 M. V. Melosi (1995), 'Equity, eco-racism and environmental history', *Environmental History Review*, 19(3): 1–16, doi:10.2307/3984909.

33 S. Fox (1981), *John Muir and His Legacy: The American Conservation Movement*, Little, Brown, Boston, MA.

34 C. Merchant (2007), *American Environmental History: An Introduction*, Columbia University Press, New York.

35 S. R. Arnstein (1969), 'A ladder of citizen participation', *Journal of the American Planning Association*, 35(4): 216–24.

36 L. Colbourne (2005), *Literature Review of Public Participation and Communicating Flood Risk*, Lindsey Colbourne Associates for ComCoast/the Environment Agency; L. Colbourne (2009), *Mainstreaming Collaboration with Communities and Stakeholders for FCERM. Improving Institutional and Social Responses to Flooding*, Joint Defra/Environment Agency Flood and Coastal Erosion Risk Management R&D Programme, Science Report: SC060019, Work Package 4, Environment Agency, Bristol, publications.environment-agency.gov.uk/pdf/SCHO0509BQBR-E-E.pdf.

37 S. R. Baillie, H. Q. P. Crick, B. E. Balmer, L. P. Beaven, I. S. Downie, S. N. Freeman, D. I. Leech, J. H. Marchant, D. G. Noble, M. J. Raven, A. P. Simpkin, R. M. Thewlsi and C. V. Wernham (2001), *Breeding Birds in the Wider Countryside: Their conservation status 2001*, BTO Research Report no. 278, British Trust for Ornithology, Thetford; R. D. Gregory,

D. G. Noble and J. Custance (2004), 'The state of play of farmland birds: population trends and conservation status of lowland farmland birds in the United Kingdom', *Ibis*, 146: 1–13.

38 Institution of Civil Engineers (2001), *Learning to Live with Rivers*, Institution of Civil Engineers, London, www.ice.org.uk/rtfpdf/ICEFlooding.pdf.

39 Defra (2002), *Farming and Food's Contribution to Sustainable Development. Economic and Statistical Analysis*, Department for Environment, Food and Rural Affairs, London.

40 M. Pitt (2007) *The Pitt Review: Lessons Learned from the 2007 Floods*, Cabinet Office, London, www.cabinetoffice.gov.uk/thepittreview/final_report.aspx.

41 M. Everard (1997), 'Floodplain protection: challenges for the next millennium', in R. G. Bailey, P. V. José and B. R. Sherwood (eds), *United Kingdom Floodplains*, Westbury Academic and Scientific Publishing, West Yorkshire, pp. 477–83.

42 Scrase. and Sheate (2005), 'Re-framing flood control'.

43 P. Opdam and D. Wascher (2004), 'Climate change meets habitat fragmentation: linking landscape and biogeographical scale levels in research and conservation', *Biological Conservation*, 117(3): 285–97.

44 J. H. Lawton, P. N. M. Brotherton, V. K. Brown, C. Elphick, A. H. Fitter, J. Forshaw, R. W. Haddow, S. Hilborne, R. N. Leafe, G. M. Mace, M. P. Southgate, W. J. Sutherland, T. E. Tew, J. Varley and G. R. Wynne (2010), *Making Space for Nature: A review of England's wildlife sites and ecological network*, Report to the Department for Environment, Food and Rural Affairs (Defra), London, www.defra.gov.uk/environment/biodiversity/documents/201009space-for-nature.pdf.

45 M. Everard and G. Kataria (2010), 'Recreational angling markets to advance the conservation of a reach of the Western Ramganga River', *Aquatic Conservation*, doi: 10.1002/aqc.1159.

46 F. Schiemer (2000), 'Fish as indicators for the assessment of the ecological integrity of large rivers', *Hydrobiologia*, 422/423: 271–8; N. Jepsen and D. Pont (eds) (2007) 'Intercalibration of fish-based methods to evaluate river ecological quality. Report from an EU inter-calibration pilot exercise', EU Joint Research Centre 8040, European Communities, Luxembourg.

47 D. Ison, N. Röling and D. Watson (2007), 'Challenges to science and society in the sustainable management and use of water: investigating the role of social learning', *Environmental Science and Policy*, 10: 499–511.

48 Millennium Ecosystem Assessment (2005), *Ecosystems and Human Well-Being*.

49 J. Porritt (2000), *Playing Safe: Science and the environment*, Thames and Hudson, London.

50 S. O. Funtowicz and J. R. Ravetz (1990), *Uncertainty and Quality in Science for Policy*, Kluwer Academic Publishers, the Netherlands; S. O. Funtowicz and J. R. Ravetz (1992), 'Three types of risk assessment and the emergence of post-normal science', in S. Krimsky and D. Golding (eds), *Social Theories of Risk*, Praeger, Westport, CT, pp. 251–74.

51 M. Elvin (1972), 'The high-level equilibrium trap: the causes of the decline of invention in the traditional Chinese textile

industries', in W. E. Willmott (ed.), *Economic Organization in Chinese Society*, Stanford University Press, Stanford, CA, pp. 137–72.

52 P. Checkland (1981), *Systems Thinking, Systems Practice*, Wiley, Chichester.

53 Everard and Powell, 'Rivers as living systems'.

54 M. Everard (2004), 'Investing in sustainable catchments', *Science of the Total Environment*, 324(1–3): 1–24.

55 Millennium Ecosystem Assessment (2005), *Ecosystems and Human Well-Being: Synthesis*.

56 Everard and Powell (2002), 'Rivers as living systems'; Ison et al. (2007), 'Challenges to science and society'.

57 Defra (2005), *Making Space for Water: Taking forward a new Government strategy for flood and coastal erosion risk management in England*, Department for Environment, Food and Rural Affairs, London.

58 Defra (2010), *Flood and Water Management Act 2010*, Department for Environment, Food and Rural Affairs, London, www.defra.gov.uk/environment /flooding/policy/fwmb.

59 R. D. Knabb, J. R. Rhome and D. P. Brown (2006), *Tropical Cyclone Report: Hurricane Katrina: 23–30 August 2005*, National Hurricane Center, www.nhc.noaa.gov/pdf/TCR-AL122005_Katrina.pdf, accessed 31 August 2010.

60 Ramsar Convention (1971), *Convention on Wetlands of International Importance Especially as Waterfowl Habitat*, 2 February, Ramsar, Iran.

61 IUCN/UNEP/WWF (1980), *World Conservation Strategy: Living Resource Conservation for Sustainable Development*, IUCN, Gland, Switzerland.

62 WCED (1987), *Our Common Future*, Oxford University Press, Oxford.

63 S. R. Brechin, P. R. Wils-Husen, C. L. Fortwangler and P. C. West (2003), *Contested Nature: Promoting International Biodiversity with Social Justice in the Twenty-First Century*, State University of New York, Albany.

64 M. Everard (2009), *The Business of Biodiversity*, WIT Press, Ashurst.

65 Everard et al. (2009), 'Flood management'.

66 M. Everard (2008), *PVC: Reaching for sustainability*, IOM3 and The Natural Step, London.

67 I. Prigogine (1997), *The End of Certainty. Time, Chaos, and the New Laws of Nature*, Free Press, New York.

5 A landscape at our service

1 G. C. Daily (1997), *Nature's Services: Societal Dependence on Natural Ecosystems*, Island Press, Washington, DC.

2 P. J. Dugan (1990), *Wetland Conservation: A Review of Current Issues and Required Action*, IUCN, Gland, Switzerland.

3 K. Annan (2001), *We the Peoples: The Role of the United Nations in the 21st Century*, United Nations, www.un.org/millennium/sg/report/full.htm.

4 Millennium Ecosystem Assessment, www.millenniumassessment.org/en/index.aspx.

5 Millennium Ecosystem Assessment (2005), *Ecosystems and Human Well-Being: Synthesis*, Island Press, Washington, DC.

6 M. Everard, J. D. Colvin, M. Mander, C. Dickens and S. Chimbuya (2009), 'Integrated catchment value systems', *Journal of Water Resource and Protection*, 3: 174–87.

7 M. Mander (2003), *Thukela Water Project: Reserve Determination Module. Part 1. IFR Scenarios in the Thukela River Catchment: Economic Impacts on Ecosystem Services*, Institute of Natural Resources, Scottsville, South Africa.

8 R. S. de Groot (1992), *Functions of Nature: Evaluation of Nature in Environmental Planning, Management, and Decision Making*, Wolters-Noordhoff, Groningen; D. U. Hooper, F. S. Chapin III, J. J. Ewel, A. Hector, P. Inchausti, S. Lavorel, J. H. Lawton, D. M. Lodge, M. Loreau, S. Naeem, B. Schmid, H. Setälä, A. J. Symstad, J. Vandermeer and D. A. Wardle (2005), 'Effects of biodiversity on ecosystem functioning: a consensus of current knowledge', *Ecological Monographs*, 75(1): 3–35.

9 J. Lawton (2010), *Making Space for Nature: A Review of England's Wildlife Sites and Ecological Network*, Department for Environment, Food and Rural Affairs, London, www.defra.gov.uk/environment/biodiversity/documents/201009space-for-nature.pdf.

10 Millennium Ecosystem Assessment (2005), *Ecosystems and Human Well-Being: Synthesis*.

11 United Nations Millennium Development Goals, www.un.org/millenniumgoals.

12 Convention on Biological Diversity, www.cbd.int/ecosystem.

13 Published in six editions between 1798 and 1826.

14 US Department of Agriculture (2009), *Forest Land Conversion, Ecosystem Services, and Economic Issues for Policy: A Review. A Forests on the Edge Report*, General Technical Report PNW-GTR-797, USDA, August 2009, www.fs.fed.us/openspace/fote/pnw-gtr797.pdf.

15 P. Olsson, C. Folke and T.P. Hughes (2008), 'Navigating the transition to ecosystem-based management of the Great Barrier Reef, Australia', *Proceedings of the National Academy of Sciences USA*, 105: 9489–94.

16 J. Liu, S. Li, Z. Ouyang, C. Tam and X. Chen (2008), 'Ecological and socioeconomic effects of China's policies for ecosystem services', *Proceedings of the National Academy of Sciences USA*, 105: 9477–82.

17 R. W. Cowling et al. (2008), 'An operational model for mainstreaming ecosystem services for implementation', *Proceedings of the National Academy of Sciences USA*, 105: 9483–8.

18 Defra (2010), 'Towards a deeper understanding of the value of nature: encouraging an interdisciplinary approach towards evidence about the value of the natural environment', Department for Environment, Food and Rural Affairs, London, www.defra.gov.uk/environment/policy/natural-environ/documents/natures-value.pdf.

19 www.teebweb.org/.

20 www.naturalvalueinitative.org.

21 www.naturalcapitalinitiative.org.uk.

6 The great food challenge

1 T. Jones, A. Bockhorst, B. McKee and A. Ndiaye (2003), *Percentage of Food Loss in the Household. Report to the United States Department of Agriculture*, Economics Research Service from the Bureau of Applied Research in Anthropology, University of Arizona.

2 Covered in a variety of media, including, for example, www.independent.co.uk/life-style/food-and-drink/news/the-16320bn-food-mountain-britons-throw-away-half-of-the-food-produced-each-year-790318.html.

3 Japan Times Herald (2009), 'Japan's food self sufficiency', *Japan Times Herald* blogspot, 7 October, japantherald.blogspot.com/2009/10/japans-food-self-sufficiency.html.

4 statistics.defra.gov.uk/esg/datasets/selfsuff.xls.

5 Food and Agriculture Organization (1993), *The State of Food and Agriculture 1993. I. Water Resource Issues and Agriculture*, UN Food and Agriculture Organization, Rome, www.fao.org/docrep/003/t0800e/t0800e0a.htm.

6 B. Bhatia and M. Falkenmark (1992), 'Water resource policies and the urban poor: innovative approaches and policy imperatives', Background paper for the ICWE, Dublin.

7 International Food Policy Research Institute (2009), *Climate Change: Impact on Agriculture and Costs of Adaptation*, International Food Policy Research Institute, Washington, DC, www.ifpri.org/sites/default/files/publications/pr21.pdf.

8 P. J. Ericksen, J. S. I. Ingram and D. M. Liverman (2009), 'Editorial: Food security and global environmental change: emerging challenges', *Environmental Science and Policy*, 12: 373–7.

9 Defra (2010), *Food 2030*, Department for Environment, Food and Rural Affairs, London, www.defra.gov.uk/foodfarm/food/pdf/food2030strategy.pdf.

10 Defra (2006), *Food Security and the UK: An Evidence and Analysis Paper*, Department for Environment, Food and Rural Affairs, December 2006, statistics.defra.gov.uk/esg/reports/foodsecurity/foodsecurity.doc.

11 UN Food and Agriculture Organization (1996), *Rome Declaration on World Food Security*, FAO, Rome.

12 D. O'Neill (2007), 'The total external environmental costs and benefits of agriculture in the UK', Environment Agency research report, www.environment-agency.gov.uk/static/documents/Research/costs_benefitapr07_1749472.pdf.

13 Royal Society (2009), *Reaping the Benefits: Science and the Sustainable Intensification of Global Agriculture*, Royal Society, London.

14 I. R. Calder (1999), *The Blue Revolution: Land Use and Integrated Water Resources Management*, Earthscan, London.

15 Ben Phalan, Unpublished data, January 2010, presented by Andrew Balmford in his presentation 'Reconciling food production and conservation: inconvenient truths about farming and the fate of wild nature' at the Conférence Internationale de Biodiversité et Agricultures, Montpelier, 4/5 November 2008.

16 *Reductions in Emissions from Deforestation and Degradation*, www.un-redd.org.

17 M. Everard, M. Bramley, K. Tatem, T. Appleby and W. Watts (2009), 'Flood management: from defence to sustainability', *Environmental Liability*, 2: 35–49.

18 David Tilman, University of Minnesota, summing up at the iap Conference on Biodiversity, Royal Society, London, 13/14 January 2010.

19 M. Everard (2005), *Water Meadows: Living Treasures in the English Landscape*, Forrest Text, Ceredigion.

20 www.slowfood.org.uk.

21 Y. Cao, J. Elliott, D. McCracken, K. Rowe, J. Whitehead and L. Wilson (2009), 'Estimating the scale of future land management requirements for the UK', Report to the Land Use Policy

Group, December 2009, www.LUPG.
org.uk.

7 Valuing land and landscapes

1 P. Hawken, A. Lovins and
H. L. Lovins (1999), *Natural Capital-
ism: Creating the Next Industrial
Revolution*, Back Bay Books.

2 D. Pearce, A. Markandya and
E. Barbier (1989), *Blueprint for a
Green Economy*, Earthscan, London;
J. Foster (ed.) (1997), *Valuing Nature?
Economics, ethics and environment*,
Routledge, London.

3 World Commission on
Environment and Development
(1987), *Our Common Future*, Oxford
University Press, Oxford.

4 M. Everard and T. Appleby
(2009), 'Safeguarding the societal
value of land', *Environmental Law
and Management*, 21: 16–23.

5 M. Everard (2009), *The Business
of Biodiversity*, WIT Press, Ashurst.

6 L. Kochtcheeva and A. Singh
(n.d.), *An Assessment of Risks and
Threats to Human Health Associated
with the Degradation of Ecosystems*,
UNEP/Division of Environmental Information, Sioux Falls, na.unep.net/
publications/heireport.pdf; World
Resources Institute (WRI), United
Nations Development Programme
(UNDP), United Nations Environment Programme (UNEP) and World
Bank (2005), *World Resources 2005:
The Wealth of the Poor – Managing Ecosystems to Fight Poverty*,
World Resources Institute (WRI),
Washington, DC; this includes
major international development
and research programmes such as
the UK's ESPA (Ecosystem Services
for Poverty Alleviation), www.nerc.
ac.uk/research/programmes/espa/.

7 J. Porritt (2005), *Capitalism
as if the World Matters*, Earthscan,
London.

8 I. Swingland (ed.) (2002),
*Capturing Carbon and Conserving
Biodiversity: The Market Approach*,
Earthscan, London.

9 M. Everard and T. Appleby
(2008), 'Ecosystem services and the
common law: evaluating the full
scale of damages', *Environmental
Law and Management*, 20: 325–39.

10 Defra (2009), *Safeguarding Our
Soils: A Strategy for England*, Department for Environment, Food and
Rural Affairs, London, www.defra.
gov.uk/environment/quality/land/
soil/documents/soil-strategy.pdf.

11 P. Dasgupta (1982), *The
Control of Resources*, Blackwell,
Oxford.

12 www.teebweb.org.

13 R. Costanza, R. d'Arge,
R. de Groot, S. Farber, M. Grasso,
B. Hannon, K. Limburg, S. Haeem,
R. V. O'Neill, J. Paruelo, R. V. Raskin,
P. Sutton and M. van den Belt (1997),
'The value of the world's ecosystem
and natural capital', *Nature*, 387:
253–60.

14 R. K. Turner, S. Georgiou and
B. Fisher (2008), *Valuing Ecosystem
Services: The Case of Multi-functional
Wetlands*, Earthscan, London.

15 M. Everard (2009), *Ecosystem
Services Case Studies*, Environment
Agency Science Report SCHO0409
BPVM, Environment Agency, Bristol.

16 Reviewed in M. Everard and
K. Capper (2004), 'Common law and
river conservation: the case for whole
systems thinking', *Environmental
Law and Management*, 16(3): 135–44.

17 I.-M. Gren, C. Folke,
R. K. Turner and I. J. Bateman
(1994), 'Primary and secondary
values of wetland ecosystems', *Environmental and Resource Economics*, 4:
55–74; R. K. Turner (1999) 'The place
of economic values in environmental
valuation', in I. J. Bateman and

K. G. Willis (eds), *Valuing Environmental Preferences*, Oxford University Press, Oxford.

18 Defra (2007b), 'Securing a healthy natural environment: an action plan for embedding an ecosystems approach', Department for Environment, Food and Rural Affairs, London; Defra (2007a), 'An introductory guide to valuing ecosystem services', Department for Environment, Food and Rural Affairs, London; Turner et al. (2008), *Valuing Ecosystem Services*.

19 Everard (2009), *Ecosystem Services Case Studies*.

20 Defra (2007a), 'An introductory guide to valuing ecosystem services'.

21 Everard (2009), *Ecosystem Services Case Studies*.

22 M. Everard (2010), 'Ecosystem services assessment of sea trout restoration work on the River Glaven, North Norfolk', Environment Agency Evidence report SCHO0110BRTZ-E-P, Environment Agency, Bristol, publications.environment-agency.gov.uk/pdf/SCHO0110BRTZ-e-e.pdf; M. Everard and S. Jevons (2010), 'Ecosystem services assessment of buffer zone installation on the upper Bristol Avon, Wiltshire', Environment Agency Evidence report SCHO0210BRXW-e-e, Environment Agency, Bristol, publications.environment-agency.gov.uk/pdf/SCHO0210BRXW-e-e.pdf; M. Everard, L. Shuker and A. Gurnell (in press), 'The Mayes Brook restoration in Mayesbrook Park, East London: an ecosystem services assessment', Environment Agency Evidence report, Environment Agency, Bristol.

23 P. Glaves, D. Egan, K. Harrison and R. Robinson (2009), *Valuing Ecosystem Services in the East of England*, East of England Environment Forum, East of England Regional Assembly and Government Office East England.

24 EFTEC (2007), 'Flood and coastal erosion risk management: economic valuation of environmental effects', *Handbook for the Environment Agency for England and Wales*, August, Economics for the Environment Consultancy Ltd (EFTEC), London.

25 N. Kosoy and E. Corbera (2010), 'Payments for ecosystem services as commodity fetishism', *Ecological Economics*, 69: 1228–36.

26 M. Mander (2003), 'Thukela Water Project: reserve determination module. Part 1', *IFR Scenarios in the Thukela River Catchment: Economic Impacts on Ecosystem Services*, Institute of Natural Resources, Scottsville, South Africa.

27 World Commission on Dams (2000), *Dams and Development: A New Framework for Better Decision-making*, Earthscan, London.

28 M. Mander and M. Everard (2008), 'The socio-economics of river management', *Environmental Scientist*, 17(3): 31–4; Everard (2009), *The Business of Biodiversity*.

29 Defra (2007a), 'An introductory guide to valuing ecosystem services'.

30 Defra (2007b) 'Securing a healthy natural environment'.

31 EFTEC (2007), *Policy Appraisal and the Environment: An Introduction to the Valuation of Ecosystem Services – Wareham Managed Realignment Case Study*, Report prepared for the Environment Agency.

32 Glaves et al. (2009), *Valuing Ecosystem Services in the East of England*.

33 M. Everard and G. Kataria (2010), 'The proposed Pancheshwar Dam, India/Nepal: a preliminary ecosystem services assessment of likely outcomes', IES research report,

www.ies-uk.org.uk/resources/papers/
pancheshwar_dam_report.pdf.

34 M. Everard, J. D. Colvin,
T. Appleby, W. Watts and S. Chimbuya (2009), 'Tools for the equitable
and sustainable use of the ecosystem
resources. Part 2: Internal markets
for the private and public benefits of
catchment use', *Environmental Law
and Management*, 20: 76–82.

35 Hawken et al. (1999), *Natural
Capitalism*; Porritt (2005), *Capitalism
as if the World Matters*.

36 Everard (2009), *The Business of
Biodiversity*.

8 Living landscapes

1 Millennium Ecosystem Assess-
ment (2005), *Ecosystems and Human
Well-Being: Synthesis*, Island Press,
Washington, DC.

2 The second appendix of the
book *Valuing Ecosystem Services* pro-
vides a comprehensive overview of
many of those relating to wetlands.
See R. K. Turner, S. Georgiou and
B. Fisher (2008), *Valuing Ecosystem
Services: The Case of Multi-functional
Wetlands*, Earthscan, London.

3 EFTEC (2007), Policy Appraisal
and the Environment: An Introduction to the Valuation of Ecosystem
Services. Wareham Managed Realignment Case Study, Defra, London.

4 Defra (2007), 'An introductory
guide to valuing ecosystem services',
Department for Environment, Food
and Rural Affairs, London.

5 M. Everard (2009), *Ecosystem
Services Case Studies*, Environment
Agency Science Report SCHO0409B-
PVM, Bristol.

6 D. Curry (2002), *The Future
of Food and Farming*, Department
for Environment, Food and Rural
Affairs, London.

7 This is reviewed in greater
detail in M. Everard (2009), *The

Business of Biodiversity*, WIT Press,
Ashurst.

8 G. Daily and K. Ellison (2004),
The New Economy of Nature, Island
Press, Washington DC (detail on
p. 63).

9 S. Wunder (2005), *Payments for
Environmental Services: Some nuts
and bolts*, CIFOR Occasional Paper
no. 42, Center for International
Forestry Research, Bogor, Indonesia.

10 S. Pagiola (2008), 'Payments
for environmental services in Costa
Rica', *Ecological Economics*, 65(4):
712–24; Daily and Ellison (2002),
*The New Economy of Nature and the
Marketplace*; M. Jenkins, S. Scherr
and M. Inbar (2004), 'Markets for
biodiversity services', *Environment*,
46(6): 32–42.

11 OECD (2010), *Paying
for Biodiversity: Enhancing the
Cost-Effectiveness of Payments for
Ecosystem Services*, OECD Publishing,
doi: 10.1787/9789264090279-en.

12 M. M. Sommerville,
J. P. G. Jones and E. J. Milner-
Gulland (2009), 'A revised conceptual
framework for payments for environmental services', *Ecology and Society*,
14(2): 34, www.ecologyandsociety.
org/vol14/iss2/art34/; T. Wünscher,
S. Engel and S. Wunder (2008),
'Spatial targeting of payments for
environmental services: a tool for
boosting conservation benefits',
Ecological Economics, 65: 822–33.

13 N. Johnson, A. White and
D. Perrot-Maître (2001), *Developing
Markets for Water Services from For-
ests: Issues and Lessons for Innovators*,
Forest Trends with World Resources
Institute and the Katoomba Group,
Washington, DC; S. Pagiola,
J. Bishop and N. Landell-Mills (eds)
(2002), *Selling Forest Environmental
Services: Market-Based Mechanisms
for Conservation and Development*,

Earthscan, London; R. Milne and T. A. Brown (1997), 'Carbon in the vegetation and soils of Great Britain', *Journal of Environmental Management*, 49(4): 413–33.

14 Information drawn from a review by D. Perrot-Maître (2006), *The Vittel Payments for Ecosystem Services: A 'perfect' PES case?*, International Institute for Environment and Development, London, UK; and also from C. Déprés, G. Grolleau and N. Mzoughi (2005), *Contracting for Environmental Property Rights: The case of Vittel*, Paper presented at the 99th Seminar of the European Association of Agricultural Economists, Copenhagen, 24–27 August, www.eaae2005.dk/CONTRIBUTED_PAPERS/S59_713_Mzoughi_etal.pdf#search= per cent22observatoire per cent20 environnement per cent202005 per cent20vittel per cent22.

15 M. Everard, W. Kenmir, C. Walters and E. Holt (2004), 'Upland hill farming for water, wildlife and food', *Freshwater Forum*, 21: 48–73.

16 See the website www.unitedutilites.com/scamp and also the extensive review in Everard (2009), *The Business of Biodiversity*.

17 PAA (2010), *United Utilities SCaMP Sustainable Catchment Management Programme: Monitoring Progress Report Year 4*, Penny Anderson Associates Ltd, Buxton, www.unitedutilities.com/scamp.htm, accessed 1 September 2010.

18 OECD (2010), *Paying for Biodiversity*.

19 N. Bannister, J. Mant and M. Janes (2005), *A Review of Catchment Scale River Restoration Projects in the UK*, River Restoration Centre, Silsoe.

20 www.therrc.co.uk.

21 www.salmon-trout.org.

22 www.wyeuskfoundation.org.

23 www.wrt.org.uk.

24 This is comprehensively reviewed by M. Everard (2004), 'Investing in sustainable catchments', *Science of the Total Environment*, 324(1/3): 1–24.

25 D. Hogan, E. Maltby and M. Blackwell(2000), *Tamar 2000 SUPPORT Project – Wetlands: Phase II Report (September 2000)*, Wetland Ecosystems Research Group (WERG), Royal Holloway Institute for Environmental Research.

26 E. Tusa (2000), *Evaluation of Economic Benefits of the Tamar 2000 Project, Phase II*, Report prepared for Westcountry Rivers Trust by the Wetland Ecosystems Research Group (WERG), Royal Holloway Institute for Environmental Research.

27 Everard (2009), *Ecosystem Services Case Studies*.

28 Tusa (2000), *Evaluation of Economic Benefits of the Tamar 2000 Project, Phase II*.

29 uk.chm-cbd.net/Default.aspx?page=7021.

30 M. Everard (2010), 'Ecosystem services assessment of sea trout restoration work on the River Glaven, North Norfolk', Environment Agency Evidence report SCHO0110BRTZ-E-P, Environment Agency, Bristol, publications.environment-agency.gov.uk/pdf/SCHO0110BRTZ-e-e.pdf.

31 M. Everard and S. Jevons (2010), 'Ecosystem services assessment of buffer zone installation on the upper Bristol Avon, Wiltshire', Environment Agency Evidence report SCHO0210BRXW-e-e, Environment Agency, Bristol, publications.environment-agency.gov.uk/pdf/SCHO0210BRXW-e-e.pdf.

32 M. Mander (2003), *Thukela Water Project: Reserve Determination Module. Part 1. IFR Scenarios in the*

Thukela River Catchment: Economic Impacts on Ecosystem Services, Institute of Natural Resources, Scottsville, South Africa.

33 Maloti Drakensberg Transfrontier Project (2007), *Payment for Ecosystem Services: Developing an Ecosystem Services Trading Model for the Mnweni/Cathedral Peak and Eastern Cape Drakensberg Areas*, ed. M. Mander, INR Report IR281, Development Bank of Southern Africa, Department of Water Affairs and Forestry, Department of Environment Affairs and Tourism, Ezemvelo KZN Wildlife, South Africa; J. Blignaut, M. Mander, R. Schulze, M. Horan, C. Dickens, C. Pringle, K. Mavundla, I. Mahlangu, A. Wilson, M. McKenzie and S. McKean (2010), 'Restoring and managing natural capital towards fostering economic development: evidence from the Drakensberg, South Africa', *Ecological Economics*, 69: 1313–23.

34 World Water Council (2006), 'Ecosystem and ecohydrology approaches to integrated water resources management', FT2.38, World Water Forum, Mexico, www.worldwatercouncil.org/index.php?id=1236&L=3.

35 World Bank (2007), *State and Trends of the Carbon Market 2007*, World Bank, Washington, DC.

36 J. Börner, S. Wunder, S. Wertz-Kanounnikoff, M. Rügnitz Tito, L. Pereira and N. Nascimento (2010), 'Direct conservation payments in the Brazilian Amazon: scope and equity implications', *Ecological Economics*, 69: 1272–82; T. Clements, A. John, K. Nielsen, D. An, S. Tan and E. J. Milner-Gulland (2010), 'Payments for biodiversity conservation in the context of weak institutions: comparison of three programs from Cambodia', *Ecological Economics*, 69: 1283–91.

37 A. Vatn (2010), 'An institutional analysis of payments of environmental services', *Ecological Economics*, 69: 1245–52.

38 R. B. Norgaard (2010), 'Ecosystem services: from eye-opening metaphor to complexity blinder', *Ecological Economics*, 69: 1219–27.

39 M. Everard and T. Appleby (2009), 'Safeguarding the societal value of land', *Environmental Law and Management*, 21: 16–23.

40 M. Everard, T. Appleby, J. D. Colvin, W. Watts and S. Chimbuya (2009), 'Modelling the private and public benefits of land use', *Environmental Law and Management*, 20: 70–75.

41 Defra (2007), 'An introductory guide to valuing ecosystem services'.

42 M. Everard, T. Appleby, J. D. Colvin, W. Watts and S. Chimbuya (2009), 'Internal markets for the private and public benefits of catchment use', *Environmental Law and Management*, 20: 76–82.

43 J. D. Colvin, F. Ballim, S. Chimbuya, M. Everard, J. Goss, G. Klarenberg, S. Ndlovu, D. Ncala and D. Weston (2009), 'Building capacity for co-operative governance as a basis for integrated water resources managing in the Inkomati and Mvoti catchments, South Africa', *Water SA*, 34(6): 681–90.

44 I. R. Calder (1999), *The Blue Revolution: Land Use and Integrated Water Resources Management*, Earthscan, London.

45 Everard et al. (2009), 'Modelling the private and public benefits of land use'.

46 Republic of South Africa (1998) National Water Act. Act no. 36 of 1998, *Government Gazette*, South Africa.

47 Calder (1999), *The Blue Revolution*.

48 For example, E. von Weizsäcker, A. B. Lovins and L. H. Lovins (1997), *Factor Four: Doubling Wealth, Halving Resource Use – a New Report to the Club of Rome*, Earthscan, London.

49 G. Murtough, B. Aretino and M. Anna (2002), *Creating Markets for Ecosystem Services (June 17, 2002)*, Productivity Commission Working Paper no. 1709, Australia, ssrn.com/abstract=322380 or doi:10.2139/ssrn.322380; UNEP (2005), 'Creating pro-poor markets for ecosystem services', Agreed notes from a high-level brainstorming workshop, 10–12 October, London, UNEP Division of Environmental Conventions, www.unep.org/dec/support/mdg_meeting_lon.htm.

9 Lessons for tomorrow's world

1 F. R. Baumgartner and B. D. Jones (1993), *Agendas and Instability in American Politics*, Chicago University Press, Chicago, IL.

2 M. Everard, M. Bramley, K. Tatem, T. Appleby and W. Watts (2009), 'Flood management: from defence to sustainability', *Environmental Liability*, 2: 35–49.

3 Millennium Ecosystem Assessment (2005), *Living Beyond Our Means: Natural Assets and Human Well-being (Statement of the MA Board)*, www.millenniumassessment.org/documents/document.429.aspx.pdf.

4 P. Johnston, M. Everard, D. Santillo and K.-H. Robèrt (2007), 'Commentaries: reclaiming the definition of sustainability', *Environmental Science and Pollution Research*, 14(1): 60–66.

5 B. Woods et al. (2007), *The SUDS Manual*, CIRIA Report C697, Construction Industry Research and Information Association, London.

6 www.greeninfrastructure.co.uk.

7 P. Roberts, C. George and J. Ravetz (2009), *Environment and the City*, Routledge, London.

8 D. J. Parker (ed.) (2000), *Floods*, Routledge, London and New York.

9 P. Taylor (1998), *An Ecological Approach to International Law: Responding to Challenges of Climate Change*, Routledge, London.

10 For example, D. Schoenbrod (2000), 'Protecting the environment in the spirit of the common law', in R. E. Meiners and A. P. Morriss (eds), *The Common Law and the Environment: Rethinking the Statutory Basis for Modern Environmental Law*, Rowman and Littlefield.

11 The Scheduled Tribes and Other Traditional Forest Dwellers (Recognition of Forest Rights) Act, 2006, is a key piece of forest legislation passed in India on 18 December 2006. It is also referred to as the 'Forest Rights Act', the 'Tribal Rights Act', the 'Tribal Bill' and the 'Tribal Land Act'.

12 M. Everard (2002), 'Implementing the Water Framework Directive: opportunities and risks', *Water Law*, 13: 30–34.

13 J. D. Colvin, F. Ballim, S. Chimbuya, M. Everard, J. Goss, G. Klarenberg, S. Ndlovu, D. Ncala and D. Weston (2009) 'Building capacity for co-operative governance as a basis for integrated water resources managing in the Inkomati and Mvoti catchments, South Africa', *Water SA*, 34(6): 681–90.

14 I. R. Calder (1999), *The Blue Revolution: Land Use and Integrated Water Resources Management*, Earthscan, London.

15 J. Morris, T. M. Hess, D. J. G. Gowing, P. B. Leeds-Harrison, N. Bannister, R. M. N. Vivash and M. Wade (2005), 'A framework

for integrating flood defence and biodiversity in washlands in England', *International Journal of River Basin Management*, 3(2): 1–11; H. Posthumus, J. R. Rouquette, J. Morris, D. J. G. Gowing and T. M. Hess (2010), 'A framework for the assessment of ecosystem goods and services; a case study on lowland floodplains in England', *Ecological Economics*, 69(7): 1510–23, doi: dx.doi.org/doi:10.1016/j.ecolecon.2010.02.011.

16 L. Colbourne (2005), *Literature Review of Public Participation and Communicating Flood Risk*, Lindsey Colbourne Associates for ComCoast/Environment Agency; L. Colbourne (2009), *Mainstreaming Collaboration with Communities and Stakeholders for FCERM*, Improving Institutional and Social Responses to Flooding, Joint Defra/Environment Agency Flood and Coastal Erosion Risk Managment R&D Programme, Science Report: SC060019, Work Package 4, Environment Agency, Bristol, publications.environment-agency.gov.uk/pdf/SCHO0509BQBR-E-E.pdf.

17 Environment Agency (2007), *Catchment Flood Management Plan Consultation – Thames Region*, Environment Agency, Reading, www.environment-agency.gov.uk/research/library/consultations/54510.aspx.

18 Environment Agency (2009), *Flood and Coastal Risk Management Strategy 2009–15*, Environment Agency, Bristol.

19 Everard et al. (2009), 'Flood management: from defence to sustainability'.

20 Environment Agency (2008), *Lower Derwent Flood Risk Management Strategy*, Environment Agency Midlands Region, Solihull.

21 N. Diederichs, T. Markewicz,

M. Mander, A. Martens and S. Zama Ngubane (2002), *eThekwini Catchments: A Strategic Tool for Management*, eThekwini Municipality, South Africa.

22 M. Everard and A. Powell (2002), 'Rivers as living systems', *Aquatic Conservation*, 12: 329–37.

23 M. Everard (2004), 'Investing in sustainable catchments', *Science of the Total Environment*, 324(1–3): 1–24.

24 J. Lawton (2010), *Making Space for Nature: A Review of England's Wildlife Sites and Ecological Network*, Department for Environment, Food and Rural Affairs, London, www.defra.gov.uk/environment/biodiversity/documents/201009space-for-nature.pdf.

25 J. Hutton and N. Leader-Williams (2003), 'Sustainable use and incentive-driven conservation: realigning human and conservation interests', *Oryx*, 37(2): 215–26.

26 R. S. de Groot, M. A. Wilson and R. M. Boumans (2002), 'A typology for the classification, description and valuation of ecosystem functions, goods and services', *Ecological Economics*, 41: 393–408.

27 C. Soulsby (n.d.), *Managing River Habitats for Fisheries: A guide to best practice*, Scottish Environment Protection Agency, Stirling.

28 www.wildtrout.org.

29 www.salmon-trout.org.

30 www.associationofriverstrusts.org.uk.

31 N. Giles (2005), *The Nature of Trout*, Perca Press, Dorset.

32 I. G. Cowx and M. J. Collares-Periera (2004), 'The role of catchment scale environmental management in freshwater fish conservation', *Fisheries Management and Ecology*, 11: 303–12.

33 As defined by M. Everard and G. Kataria (forthcoming), 'Recrea-

tional angling markets to advance the conservation of a reach of the Western Ramganga River', *Aquatic Conservation*.

34 www.merseybasin.org.uk.

35 G. Petts, J. Heathcote and D. Martin (2001), *Urban Rivers: Our Inheritance and Future*, IWA Publishing, London.

36 K. Nam-choon (2005), 'Ecological restoration and revegetation works in Korea', *Landscape and Ecological Engineering*, 1: 77–83.

37 CABE 2010 Cheonggyecheon Restoration Project, Seoul, South Korea, www.cabe.org.uk/case-studies/cheonggyecheon-restoration-project, accessed 26 July 2010.

38 Environment Agency (2007), *Catchment Flood Management Plan Consultation – Thames Region*.

39 M. Everard, J. D. Colvin, M. Mander, C. Dickens and S. Chimbuya (2009), 'Integrated catchment value systems', *Journal of Water Resource and Protection*, 3: 174–87.

40 J. Igoe and T. Kelsall (2005), 'Introduction: between a rock and a hard place', in J. Igoe and T. Kelsall, *African NGOs, Donors, and the State: Between a rock and a hard place*, Carolina Academic Press, Durham, NC, pp. 1–33.

41 www.wessexwater.co.uk.

42 M. Everard (2009), *The Business of Biodiversity*, WIT Press, Ashurst.

43 P. Carty and S. Payne (1998), *Angling and the Law*, Merlin Unwin Books, Ludlow.

44 M. Everard (2005), *Water Meadows: Living Treasures in the English Landscape*, Forrest Text, Ceredigion.

10 The people's land

1 Southern African Sub-Global Assessment (SAfMA), www.millenniumassessment.org/en/SGA.Safma.aspx.

2 Jakarta Bay and Bunaken Ecosystems Sub-Global Assessment, www.millenniumassessment.org/en/SGA.Indonesia.aspx.

3 Department of Health (1998), 'Communicating about risks to public health – pointers to good practice', Department of Health, London.

4 E. O. Wilson (1990), *Sociobiology*, Harvard University Press, Boston, MA; E. O. Wilson (1999), *Consilience*, Abacus, London.

5 T. Roszak, M. E. Gomes and A. D. Kanner (1995), *Ecopsychology: Restoring the Earth, Healing the Mind*, Sierra Book Club, San Francisco.

6 R. S. Ulrich (1984), 'View through a window may influence recovery from surgery', *Science*, 224: 420–21.

7 D. Wanless (2002), *Securing Our Future Health: Taking a Long-Term View. Final Report (April 2002)*, HM Treasury, London, www.hm-treasury.gov.uk/Consultations_and_Legislation/wanless/consult_wanless_final.cfm.

8 J. Pretty, M. Griffin, J. Peacock, R. Hine, M. Sellens and N. South (2005), 'A countryside for health and wellbeing: the physical and mental health benefits of green exercise – executive summary', *Countryside Recreation*, 13: 2–7.

9 DETR (1999), 'Quality of life counts: indicators for a strategy for sustainable development for the United Kingdom – a baseline assessment', Department of Transport, Environment and the Regions, London.

10 DETR (1999), *A Better Quality of Life: A strategy for sustainable development in the UK*, Cm 4345, May, Department of Transport,

Environment and the Regions, London; Defra (2005) *Securing the Future – UK Government sustainable development strategy*, Cm 6467, Department for the Environment, Food and Rural Affairs, London, www.sustainable-development.gov.uk.

11 G. Walker, J. Fairburn, G. Smith and G. Mitchell (2003), *Environmental Quality and Social Deprivation*, R&D Technical Report. Environment Agency, Bristol; H. Chalmers and J. Colvin (2005), 'Addressing environmental inequalities in UK policy: an action research perspective', *Local Environment*, 10(4): 333–60.

12 A. Naess (1989), *Ecology, Community and Lifestyle*, Cambridge University Press, Cambridge.

13 *Mystique of the Earth*. Interviewed by Caroline Webb in *Caduceus* magazine, 59.

14 A. Leopold (1949), *A Sand County Almanac*, Ballantine, New York.

15 WCED (1987), *Our Common Future*, Oxford University Press, Oxford.

16 Extract from p. 60 of P. Johnston, M. Everard, D. Santillo and K.-H. Robèrt (2007), 'Reclaiming the definition of sustainability', *Environmental Science and Pollution Research*, 14(1): 60–66.

17 Extract from p. 175 of M. Everard, J. D. Colvin, M. Mander, C. Dickens and S. Chimbuya (2009), 'Integrated catchment value systems', *Journal of Water Resource and Protection (JWARP)*, 3: 174–87.

18 J. Porritt (2000), *Playing Safe: Science and the environment*, Thames and Hudson, London.

19 J. Cairns (2002) 'Goals and conditions for sustainable world: a collection of papers', *Ethics in Science and Environmental Politics*, Book 1, Inter-Research, Oldendorf/Luhe, Germany; J. Cairns (2003), 'Eco ethics and sustainability ethics, a collection of papers part 1', *Ethics in Science and Environmental Politics*, Book 2, Inter-Research, Oldendorf/Luhe, Germany; J. Cairns (2004), 'Eco ethics and sustainability ethics, a collection of papers part 2', *Ethics in Science and Environmental Politics*, Book 2, Inter-Research, Oldendorf/Luhe, Germany.

20 www.naturalstep.org.

21 Johnston et al. (2007), 'Reclaiming the definition of sustainability'.

22 M. Moore (2001), *Stupid White Men*, HarperCollins, New York.

23 H. Yin and C. Li (2001), 'Human impact on floods and flood disasters on the Yangtze River', *Geomorphology*, 41: 105–9.

24 www.naturalcapitalproject.org/ConEX/China_ConEX_Brochure_100708.pdf.

25 E. Ostrom, J. Burger, C. B. Field, R. B. Norgaard and D. Policansky (1999), 'Revisiting the commons: local lessons, global challenges', *Science*, 284: 278–82.

26 C. Burr (2009), 'Usufruct: end private property to solve the financial crisis and create food security', *Culture Change*, 3 February, culturechange.org/cms/index.php?option=com_content&task=view&id=310&Itemid=1.

27 P. Hawken, A. Lovins and H. L. Lovins (1999), *Natural Capitalism: Creating the Next Industrial Revolution*, Back Bay Books; J. Porritt (2005), *Capitalism as if the World Matters*, Earthscan, London.

28 J. Elkington and J. Hailes (1988), *The Green Consumer Guide: From Shampoo to Champagne, How to Buy Goods That Don't Cost the Earth*, Gollancz, London; J. Diamond

(2005), *Collapse: How Societies Choose to Fail or Succeed*, Viking Penguin, London.

29 M. Everard (2009), *The Business of Biodiversity*, WIT Press, Ashurst.

30 Hawken et al., *Natural Capitalism*.

31 Diamond (2005), *Collapse*.

32 Porritt (2000), *Playing Safe*.

33 A. R. Turton, H. J. Hattinh, G. A. Maree, D. J. Roux, M. Claassen and W. F. Strydom (eds) (2006), *Governance as the Trialogue: Government-society-science in Transition (Water Resources Development and Management)*, Springer.

34 N. Diederichs, T. Markewicz, M. Mander, A. Martens and S. Zama Ngubane (2002), *eThekwini Catchments: A Strategic Tool for Management*, eThekwini Municipality, South Africa.

35 Everard et al. (2009), 'Integrated catchment value systems'.

Bibliography

Allaby, M. (1998) *Oxford Dictionary of Ecology*, Oxford: Oxford University Press.

Annan, K. (2001) *We the Peoples: The Role of the United Nations in the 21st Century*, United Nations.

Armstrong, W. A. (1981) 'The influence of demographic factors on the position of the agricultural labourer in England and Wales, c. 1750–1914', *Agricultural History Review*, 29: 71–82.

Arnstein, S. R. (1969) 'A ladder of citizen participation', *Journal of the American Planning Association*, 35(4): 216–24.

Asmal, K. (2008) 'Reflections on the birth of the National Water Act, 1998', *Water SA*, 34(6): 662–4.

Baillie, S. R., H. Q. P. Crick, B. E. Balmer, L. P. Beaven, I. S. Downie, S. N. Freeman, D. I. Leech, J. H. Marchant, D. G. Noble, M. J. Raven, A. P. Simpkin, R. M. Thewlsi and C. V. Wernham (2001) 'Breeding birds in the wider countryside: their conservation status 2001', *BTO Research Report* no. 278, Thetford: BTO.

Balmford, A. and T. Whitten (2003) 'Who should pay for tropical conservation, and how could these costs be met?', *Oryx*, 37(2): 238–50.

Bannister, N., J. Mant and M. Janes (2005) 'A review of catchment scale river restoration projects in the UK', Silsoe: River Restoration Centre.

Baudry, J., I. R. Bowler, R. Kronert and A. Reenberg (eds) (1999) *Land-use Changes and Their Environmental Impact in Rural Areas in Europe (Man and the Biosphere)*, Oxford: Taylor and Francis.

Baum, R.M. (2010) 'Perspective: sustainable growth is an oxymoron', *Government and Policy*, 88(45): 44–7.

Baumgartner, F. R. and B. D. Jones (1993) *Agendas and Instability in American Politics*, Chicago, IL: Chicago University Press.

Beinart, W. and P. Coates (1995) *Environment and History: The Taming of Nature in the USA and South Africa*, London: Routledge.

Benington, J. (2010) 'From private choice to public value?', in J. Benington and M. H. Moore (2010) *Public Value: Theory and Practice*, New York: Palgrave Macmillan.

Bertalanffy, L. von. (1940) 'The organization considered as a physical system', Reprinted in *General Systems Theory: Foundations, development, applications*, New York: Braziller, pp. 120–38.

Bettey, J. (1999) 'The development of water meadows in the southern counties', in H. Cook and T. Williamson (1999), *Water Management in the English Landscape: Field, marsh and meadow*, Edinburgh: Edinburgh University Press, pp. 179–95.

Bhatia, B. and M. Falkenmark (1992) *Water Resource Policies* and the Urban Poor: Innovative approaches and policy imperatives, Background paper for the ICWE, Dublin.

Bishop, J., S. Kapila, F. Hicks, P. Mitchell and F. Vorhies (2008) 'Building biodiversity business', London and Gland, Switzerland: Shell International Limited and the International Union for Conservation of Nature, p. 164.

Blignaut, J., M. Mander, R. Schulze, M. Horan, C. Dickens, C. Pringle, K. Mavundla, I. Mahlangu, A. Wilson, M. McKenzie and S. McKean (2010) 'Restoring and managing natural capital towards fostering economic development: evidence from the Drakensberg, South Africa', Ecological Economics, 69: 1313–23.

Börner, J., S. Wunder, S. Wertz-Kanounnikoff, M. Rügnitz Tito, L. Pereira and N. Nascimento (2010) 'Direct conservation payments in the Brazilian Amazon: scope and equity implications', Ecological Economics, 69: 1272–82.

Brechin, S. R., P. R. Wils-Husen, C. L. Fortwangler and P. C. West (2003) Contested Nature: Promoting International Biodiversity with Social Justice in the Twenty-first Century, Albany: State University of New York.

Brockington, D. (2009) Celebrity and the Environment: Fame, wealth and power in conservation, London: Zed Books.

Bruns, B. (2005) 'Community-based principles for negotiating water rights: some conjectures on assumptions and priorities', International workshop on African Water Laws: Plural Legislative Frameworks for Rural Water Management in Africa, Johannesburg, 26–28 January.

Burr, C. (2009) 'Usufruct: end private property to solve the financial crisis and create food security', Culture Change, culturechange.org/cms/index.php?option=com_content&task=view&id=310&Itemid=1, accessed 3 February 2009.

Burton, I., R. W. Kates and G. F. White (1978) The Environment as Hazard, New York: Oxford University Press.

CABE (2010) 'Cheonggyecheon restoration project', Seoul, www.cabe.org.uk/case-studies/cheonggyecheon-restoration-project, accessed 26 July 2010.

Cairns, J. (2002) 'Goals and conditions for sustainable world: a collection of papers', Ethics in Science and Environmental Politics, Book 1, Oldendorf/Luhe, Germany: Inter-Research.

— (2003) 'Eco ethics and sustainability ethics: a collection of papers part 1', Ethics in Science and Environmental Politics, Book 2, Oldendorf/Luhe, Germany: Inter-Research.

Calder, I. R. (1999) The Blue Revolution: Land Use and Integrated Water Resources Management, London: Earthscan.

Cao, Y., J. Elliott, D. McCracken, K. Rowe, J. Whitehead and L. Wilson (2009) 'Estimating the scale of future land management requirements for the UK', Report to the Land Use Policy Group, www.LUPG.org.uk, accessed December 2009.

Carson, R. (1962) Silent Spring, London: Hamish Hamilton.

Carty, P. and S. Payne (1998) Angling and the Law, Ludlow: Merlin Unwin Books.

Chalmers, H. and J. Colvin (2005) 'Addressing environmental inequalities in UK policy: an action research perspective', *Local Environment*, 10(4): 333–60.

Chambers, J. D. and G. E. Mingay (1966) *The Agricultural Revolution 1750–1850*, London: Batsford.

Checkland, P. (1981) *Systems Thinking, Systems Practice*, Chichester: Wiley.

Chi, C. C. (2001) 'Capitalist expansion and indigenous land rights: emerging environmental justice issues in Taiwan', *Asia Pacific Journal of Anthropology*, 2(2): 135–53.

Clements, T., A. John, K. Nielsen, D. An, S. Tan and E.J. Milner-Gulland (2010) 'Payments for biodiversity conservation in the context of weak institutions: comparison of three programs from Cambodia', *Ecological Economics*, 69: 1283–91.

Cock, J. and E. Koch (1991) *Going Green: People, Politics, and the Environment in South Africa*, Cape Town: Oxford University Press.

Colbourne, L. (2005) 'Literature review of public participation and communicating flood risk', Lindsey Colbourne Associates for ComCoast/Environment Agency.

— (2009) *Mainstreaming Collaboration with Communities and Stakeholders for FCERM. Improving Institutional and Social Responses to Flooding*, Joint Defra/Environment Agency Flood and Coastal Erosion Risk Managment R&D Programme, Science Report: SC060019, Work Package 4, Bristol: Environment Agency.

Colvin, J., F. Ballim, S. Chimbuya, M. Everard, J. Goss, G. Klarenberg, S. Ndlovu, D. Ncala and D. Weston (2008) 'Building capacity for co-operative governance as a basis for integrated water resources managing in the Inkomati and Mvoti catchments, South Africa', *Water SA*, 34(6): 681–90.

Costanza, R., R. d'Arge, R. de Groot, S. Farber, M. Grasso, B. Hannon, K. Limburg, S. Haeem, R. V. O'Neill, J. Paruelo, R. V. Raskin, P. Sutton and M. Van den Belt (1997) 'The value of the world's ecosystem and natural capital', *Nature*, 387: 253–60.

Cowan, G. (2002) 'Nomadology in architecture: ephemerality, movement and collaboration', University of Adelaide, hdl. handle.net/2440/37830 [1].

Cowling, R. W. et al. (2008) 'An operational model for mainstreaming ecosystem services for implementation', *Proceedings of the National Academy of Sciences USA*, 105: 9483–8.

Cowx, I. G. and M. J. Collares-Periera (2004) 'The role of catchment scale environmental management in freshwater fish conservation', *Fisheries Management and Ecology*, 11: 303–12.

Cox, S. J. B. (1985) 'No tragedy on the commons', *Environmental Ethics*, 7: 49–61.

Curry, D. (2002) *Future of Food and Farming*, London: Defra (Department for Environment, Food and Rural Affairs).

Dahlman, C. J. (2008) *The Open Field System and Beyond: A Property Rights Analysis of an Economic Institution*, Cambridge: Cambridge University Press.

Daily, G. C. (1997) *Nature's Services Societal Dependence on Natural Ecosystems*, Washington, DC: Island Press.

Daily, G. C. and K. Ellison (2002) *The New Economy of Nature and the Marketplace: The Quest to Make Conservation Profitable*, Washington, DC: Island Press.

— (2004) *The New Economy of Nature*, Washington, DC: Island Press.

Dasgupta, P. (1982) *The Control of Resources*, Oxford: Blackwell.

Denny, P. (1994) 'Biodiversity and wetlands', *Wetlands Ecology and Management*, 3: 55–61.

Defra (2002) 'Farming and food's contribution to sustainable development, economic and statistical analysis', London: Department for Environment, Food and Rural Affairs.

— (2005a) 'Securing the future – UK government sustainable development strategy', Cm 6467, London: Department for Environment, Food and Rural Affairs.

— (2005b) 'Making space for water: taking forward a new government strategy for flood and coastal erosion risk management in England', First government response to the autumn 2004 floods, London: Department for Environment, Food and Rural Affairs.

— (2006) 'Food security and the UK: an evidence and analysis paper', London: Department for Environment, Food and Rural Affairs, statistics.defra.gov.uk/esg/reports/foodsecurity/foodsecurity.doc.

— (2007a) 'An introductory guide to valuing ecosystem services', London: Department for Environment, Food and Rural Affairs.

— (2007b) 'Securing a healthy natural environment: an action plan for embedding an ecosystems approach', London: Department for Environment, Food and Rural Affairs.

— (2009) 'Safeguarding our soils: a strategy for England', London: Department for Environment, Food and Rural Affairs, www.defra.gov.uk/environment/quality/land/soil/documents/soil-strategy.pdf.

— (2010a) 'Flood and Water Management Act 2010', London: Department for Environment, Food and Rural Affairs, www.defra.gov.uk/environment/flooding/policy/fwmb.

— (2010b) 'Food 2030', London: Department for Environment, Food and Rural Affairs, www.defra.gov.uk/foodfarm/food/pdf/food2030strategy.pdf.

— (2010c) 'Towards a deeper understanding of the value of nature: encouraging an interdisciplinary approach towards evidence about the value of the natural environment', London: Department for Environment, Food and Rural Affairs, www.defra.gov.uk/environment/policy/natural-environ/documents/natures-value.pdf.

De Groot, R. S. (1992) *Functions of Nature: Evaluation of Nature in Environmental Planning, Management, and Decision Making*, Groningen: Wolters-Noordhoff.

De Groot, R. S., M. A. Wilson and R. M. Boumans (2002) 'A typology for the classification, description and valuation of ecosystem functions, goods and services', *Ecological Economics*, 41: 393–408.

Department of Health (1998) 'Communicating about risks to public health – pointers to good practice', London: Department of Health.

Déprés, C., G. Grolleau and N. Mzoughi (2005) *Contracting for Environmental Property Rights: The case of Vittel*, Paper presented at the 99th Seminar of the European Association of Agricultural Economists, Copenhagen, 24–27 August, www.

eaae2005.dk/CONTRIBUTED_
PAPERS/S59_713_Mzoughi_etal.
pdf#search=%22observatoire%
20environnement%202005%20
vittel%22.

DETR (1999a) 'A better quality of life:
a strategy for sustainable development in the UK', Cm 4345, May,
London: Department of Transport,
Environment and the Regions.

— (1999b) 'Quality of life counts:
indicators for a strategy for
sustainable development for
the United Kingdom, a baseline
assessment', London: Department of Transport, Environment
and the Regions.

Devall, W. and G. Sessions (2001)
*Deep Ecology: Living as if Nature
Mattered*, Layton: Gibbs Smith.

Diamond, J. (1998) *Guns, Germs and
Steel*, New York: Viking Penguin.

— (2005) *Collapse: How Societies
Choose to Fail or Succeed*, New
York: Viking Penguin.

Diederichs, N., T. Markewicz,
M. Mander, A. Martens and
S. Zama Ngubane (2002)
*eThekwini Catchments: A Strategic
Tool for Management*, eThekwini
Municipality, South Africa.

Dietz, T., E. Ostrom and P. C. Stern
(2003) 'The struggle to govern the
commons', *Science*, 302: 1907–12.

Dugan, P. J. (1990) *Wetland Conservation: A Review of Current Issues
and Required Action*, Gland,
Switzerland: IUCN.

EFTEC (2007a) 'Policy appraisal and
the environment: an introduction
to the valuation of ecosystem
services – Wareham managed
realignment case study', Report
prepared for the Environment
Agency.

— (2007b) 'Flood and coastal
erosion risk management: economic valuation of environmental

effects', 15500, Handbook for the
Environment Agency for England
and Wales, August, London:
Economics for the Environment
Consultancy Ltd.

Ehrlich, P. and A. Ehrlich (1982)
*Extinction: The Causes and
Consequences of the Disappearance
of Species*, London: Gollancz.

Elkington, J. and J. Hailes (1988)
*The Green Consumer Guide: From
Shampoo to Champagne, How to
Buy Goods That Don't Cost the
Earth*, London: Gollancz.

Elvin, M. (1972) 'The high-level
equilibrium trap: the causes of
the decline of invention in the
traditional Chinese textile industries', in W. E. Willmott (ed.),
*Economic Organization in Chinese
Society*, Stanford, CA: Stanford
University Press, pp. 137–72.

Engels, F. (1882) *Die Entwicklung des
Sozialismus von der Utopie zur
Wissenschaft*, Zurich: Hottingen.

English Nature (2001) *Accentuate the
Positive – funding positive management on SSSIs*, Press release,
2 March.

Environment Agency (2007) 'Catch-
ment Flood Management Plan
Consultation – Thames Region',
Reading: Environment Agency.

— (2008) 'Lower Derwent Flood Risk
Management Strategy', Solihull:
Environment Agency Midlands
Region.

— (2009) 'Flood and Coastal Risk
Management Strategy 2009–15',
Bristol: Environment Agency.

Environmental Audit Committee
(2006) 'Outflanked: the World
Trade Organisation, international
trade and sustainable development', HC 1455, Eleventh Report
of Session 2005–06.

Ericksen, P. J., J. S. I. Ingram and
D. M. Liverman (2009) 'Edit-

orial: food security and global environmental change: emerging challenges', *Environmental Science and Policy*, 12: 373–7.

Everard, M. (1997) 'Floodplain protection: challenges for the next millennium', in R. G. Bailey, P. V. José and B. R. Sherwood (eds), *United Kingdom Floodplains*, West Yorkshire: Westbury Academic and Scientific Publishing, pp. 477–83.

— (2002) 'Implementing the Water Framework Directive: opportunities and risks', *Water Law*, 13: 30–34.

— (2004) 'Investing in sustainable catchments', *Science of the Total Environment*, 324(1–3): 1–24.

— (2005) *Water Meadows: Living treasures in the English landscape*, Ceredigion: Forrest Text.

— (2008) 'PVC: reaching for sustainability', London: IOM3 and The Natural Step.

— (2009) *The Business of Biodiversity*, Ashurst: WIT Press.

— (2010) 'Ecosystem services assessment of sea trout restoration work on the River Glaven, North Norfolk', Environment Agency Evidence report, Bristol: Environment Agency.

Everard, M. and T. Appleby (2008) 'Ecosystem services and the common law: evaluating the full scale of damages', *Environmental Law and Management*, 20: 325–39.

— (2009) 'Safeguarding the societal value of land', *Environmental Law and Management*, 21: 16–23.

Everard, M. and K. Capper (2004) 'Common law and river conservation: the case for whole systems thinking', *Environmental Law and Management*, 16(3): 135–44.

Everard, M. and S. Jevons (2010) 'Ecosystem services assessment of buffer zone installation on the upper Bristol Avon, Wiltshire', Environment Agency Evidence report, Bristol: Environment Agency.

Everard, M. and G. Kataria (2010a) 'Recreational angling markets to advance the conservation of a reach of the Western Ramganga River', *Aquatic Conservation*, doi: 10.1002/aqc.1159.

— (2010b) 'The proposed Pancheshwar Dam, India/Nepal: a preliminary ecosystem services assessment of likely outcomes', IES research report, www. ies-uk.org.uk/resources/papers/ pancheshwar_dam_report.pdf.

Everard, M. and H. L. Moggridge (forthcoming) 'Rediscovering the value of urban rivers', *Urban Ecosystems*.

Everard, M. and A. Powell (2002) 'Rivers as living systems', *Aquatic Conservation*, 12: 329–37.

Everard, M., T. Appleby, J. D. Colvin, W. Watts and S. Chimbuya (2009a) 'Internal markets for the private and public benefits of catchment use', *Environmental Law and Management*, 20: 76–82.

— (2009b) 'Modelling the private and public benefits of land use', *Environmental Law and Management*, 20: 70–75.

Everard, M., M. Bramley, K. Tatem, T. Appleby and W. Watts (2009) 'Flood management: from defence to sustainability', *Environmental Liability*, 2: 35–49.

Everard, M., J. D. Colvin, T. Appleby, W. Watts and S. Chimbuya (2009) 'Tools for the equitable and sustainable use of the ecosystem resources. Part 2: Internal markets for the private and public benefits of catchment use', *Environmental Law and Management*, 20: 76–82.

Everard, M., J. D. Colvin, M. Mander, C. Dickens and S. Chimbuya (2009) 'Integrated catchment value systems', *Journal of Water Resource and Protection*, 3: 174–87.

Everard, M., W. Kenmir, C. Walters and E. Holt (2004) 'Upland hill farming for water, wildlife and food', *Freshwater Forum*, 21: 48–73.

Everard, M., L. Shuker and A. Gurnell (forthcoming) 'The Mayes Brook restoration in Mayesbrook Park, East London: an ecosystem services assessment', Environment Agency Evidence report, Bristol: Environment Agency.

Food and Agriculture Organization (1993) 'The state of food and agriculture 1993', I, Water Resource Issues and Agriculture, Rome: UN Food and Agriculture Organization, www.fao.org/docrep/003/t0800e/t0800e0a.htm.

Foster, J. (ed.) (1997) *Valuing Nature? Economics, ethics and environment*, London: Routledge.

Fox, S. (1981) *John Muir and His Legacy: The American Conservation Movement*, Boston, MA: Little, Brown.

Funtowicz, S. O. and J. R. Ravetz (1990) *Uncertainty and Quality in Science for Policy*, The Netherlands: Kluwer Academic.

— (1992) 'Three types of risk assessment and the emergence of post-normal science', in S. Krimsky and D. Golding (eds), *Social Theories of Risk*, Westport, CT: Praeger, pp. 251–74.

Ghimire, K. B. and M. P. Pimbert (1997) *Social Change and Conservation*, London: Earthscan.

Giddens, A. (2006) *Sociology*, Cambridge: Polity Press.

Giles, N. (2005) *The Nature of Trout*, Dorset: Perca Press.

Glaves, P., D. Egan, K. Harrison and R. Robinson (2009) *Valuing Ecosystem Services in the East of England*, East of England Environment Forum, East of England Regional Assembly and Government Office East England.

Gregory, R. D., D. G. Noble and J. Custance (2004) 'The state of play of farmland birds population trends and conservation status of lowland farmland birds in the United Kingdom', *Ibis*, 146: 1–13.

Gren, I. M., C. Folke, R. K. Turner and I. J. Bateman (1994) 'Primary and secondary values of wetland ecosystems', *Environmental and Resource Economics*, 4: 55–74.

Hammond, J. L. and B. Hammond (1911) *The Village Labourer, 1760–1832: A Study in the Government of England before the Reform Bill*, London: Longmans, Green and Co.

— (1917) *The Town Labourer*, London: Longmans, Green and Co.

Hardin, G. (1968) 'The tragedy of the commons', *Science*, 162: 1243–8.

Hawken, P. (1993) *The Ecology of Commerce: A Declaration of Sustainability*, New York: HarperCollins.

Hawken, P., A. Lovins and H. L. Lovins (1999) *Natural Capitalism: Creating the Next Industrial Revolution*, New York: Back Bay Books.

Hogan, D., E. Maltby and M. Blackwell (2000) 'Tamar 2000 SUPPORT Project – Wetlands: Phase II Report (September 2000)', Wetland Ecosystems Research Group (WERG), Royal Holloway Institute for Environmental Research.

Hooper, D. U., F. S. Chapin III, J. J. Ewel, A. Hector, P. Inchausti, S. Lavorel, J. H. Lawton,

D. M. Lodge, M. Loreau, S. Naeem, B. Schmid, H. Setälä, A. J. Symstad, J. Vandermeer and D. A. Wardle (2005) 'Effects of biodiversity on ecosystem functioning: a consensus of current knowledge', *Ecological Monographs*, 75(1): 3–35.

Howarth, W. (2001) 'Water, water everywhere', *Water Law*, 12: 1–3.

— (2002) 'Editorial: implementing flood protection policy through planning', *Water Law*, 13: 247–8.

Hrezo, M. S. and W. E. Hrezo (1985) 'From antagonistic to co-operative federalism in water resources development: a model for reconciling federal, state and local programs, policies and planning', *American Journal of Economics and Sociology*, 44(2).

Hutton, J. and N. Leader-Williams (2003) 'Sustainable use and incentive-driven conservation: realigning human and conservation interests', *Oryx*, 37(2): 215–26.

Igoe, J. and B. Croucher (2007) 'Poverty alleviation meets the spectacle of nature: does reality matter?', *Conservation and Society*, 5(4): 534–61.

Igoe, J. and T. Kelsall (2005) *African NGOs, Donors, and the State: Between a rock and a hard place*, Durham, NC: Carolina Academic Press.

Institution of Civil Engineers (2001) 'Living with rivers', London: Institution of Civil Engineers, www.ice.org.uk/rtfpdf/ICEFlooding.pdf.

International Food Policy Research Institute (2009) 'Climate change: impact on agriculture and costs of adaptation', Washington, DC: International Food Policy Research Institute, www.ifpri.org/sites/default/files/publications/pr21.pdf.

Ison, D., N. Röling and D. Watson (2007) 'Challenges to science and society in the sustainable management and use of water: investigating the role of social learning', *Environmental Science and Policy*, 10: 499–511.

IUCN/UNEP/WWF (1980) 'World Conservation Strategy: living resource conservation for sustainable development', Jakarta Bay and Bunaken Ecosystems Sub-Global Assessment, Gland, Switzerland: IUCN, www.millenniumassessment.org/en/SGA.Indonesia.aspx.

Japan Times Herald (2009) 'Japan's food self sufficiency', *Japan Times Herald*, 7 October.

Jenkins, M., S. Scherr and M. Inbar (2004) 'Markets for biodiversity services', *Environment*, 46(6): 32–42.

Jepsen, N. and D. Pont (eds) (2007) 'Intercalibration of fish-based methods to evaluate river ecological quality', Report from an EU intercalibration pilot exercise, Luxembourg: EU Joint Research Centre 8040, European Communities.

Johnson, C., S. Tunstall and E. Penning-Rowsell (2004) 'Crises as catalysts for adaptation: human response to major floods', Publication no. 511, Flood Hazard Research Centre.

Johnson, N., A. White and D. Perrot-Maître (2001) 'Developing markets for water services from forests: issues and lessons for innovators', Washington, DC: Forest Trends with World Resources Institute and the Katoomba Group.

Johnston, P., M. Everard, D. Santillo and K. H. Robèrt (2007) 'Reclaiming the definition of sustainability', *Environmental Science and Pollution Research*, 14(1): 60–66.

Jones, T., A. Bockhorst, B. McKee and A. Ndiaye (2003) 'Percentage of food loss in the household', Report to the United States Department of Agriculture, Economics Research Service from the Bureau of Applied Research in Anthropology, University of Arizona.

Kameri-Mbote, P. and P. Cullet (1996) 'Environmental justice and sustainable development: integrating local communities in environmental management', IELRC Working Paper 1996 – 1, *Environmental History Review*, 19(3): 1–16.

Khatibi, R. H. (2002) 'Systemic knowledge management in hydraulic systems – Paper 1: a postulate on paradigm shifts as a methodological tool', *Journal of Hydroinformatics*, 5(2): 127–40.

Knabb, R. D., J. R. Rhome and D. P. Brown (2006) 'Tropical cyclone report: Hurricane Katrina: 23–30 August 2005', National Hurricane Center, www.nhc.noaa.gov/pdf/TCR-AL122005_Katrina.pdf.

Kochtcheeva, L. and A. Singh (n.d.) 'An assessment of risks and threats to human health associated with the degradation of ecosystems', Sioux Falls: UNEP/Division of Environmental Information, na.unep.net/publications/heireport.pdf.

Kosoy, N. and E. Corbera (2010) 'Payments for ecosystem services as commodity fetishism', *Ecological Economics*, 69: 1228–36.

Kottelat, M. and J. Freyhof (2007) *Handbook of European Freshwater Fishes*, Cornol: Publications Kottelat.

Kuhn, T. S. (1962) *The Structure of Scientific Revolutions*, Chicago, IL: University of Chicago Press.

Lawton, J. (2010) *Making Space for Nature: A Review of England's Wildlife Sites and Ecological Network*, London: Department for Environment, Food and Rural Affairs.

Lawton, J. H., P. N. M. Brotherton, V. K. Brown, C. Elphick, A. H. Fitter, J. Forshaw, R. W. Haddow, S. Hilborne, R. N. Leafe, G. M. Mace, M. P. Southgate, W. J. Sutherland, T. E. Tew, J. Varley and G. R. Wynne (2010) *Making Space for Nature: A review of England's Wildlife Sites and Ecological Network*, Report to the Department for Environment, Food and Rural Affairs, London.

Leopold, A. (1949) *A Sand County Almanac: And Essays on Conservation from Round River*, New York: Oxford University Press.

Liu, J., S. Li, Z. Ouyang, C. Tam and X. Chen (2008) 'Ecological and socioeconomic effects of China's policies for ecosystem services', *Proceedings of the National Academy of Sciences USA*, 105: 9477–82.

Mander, M. (2003) 'Thukela Water Project: Reserve Determination Module. Part 1. IFR scenarios in the Thukela river catchment: economic impacts on ecosystem services', Scottsville, South Africa: Institute of Natural Resources.

— (ed.) (2007) 'Payment for ecosystem services: developing an ecosystem services trading model for the Mnweni/Cathedral Peak and Eastern Cape Drakensberg areas', INR Report IR281, Maloti Drakensberg Transfrontier Project, Development Bank of Southern Africa, Department of Water Affairs and Forestry,

Department of Environment Affairs and Tourism, Ezemvelo KZN Wildlife, South Africa.

Mander, M. and M. Everard (2008) 'The socio-economics of river management', *Environmental Scientist*, 17(3): 31–4.

Marsh, G. P. (1864) *Man and Nature; or, physical geography as modified by human action*, Whitefish, MT: Kessinger Publishing.

Medawar, P. B. and J. S. Medawar (1977) *The Life Science*, London: Wildwood House.

Meiners, R. E. and A. P. Morriss (eds) (2000) *The Common Law and the Environment: Rethinking the Statutory Basis for Modern Environmental Law*, Lanham, MD: Rowman and Littlefield.

Melosi, M.V. (1995) 'Equity, eco-racism and environmental history', *Environmental History Review*, 19(3): 1–16, doi:10.2307/3984909.

Merchant, C. (2007) *American Environmental History: An Introduction*, New York: Columbia University Press.

Millennium Ecosystem Assessment (2004) *Millennium Ecosystem Assessment*, Washington, DC: Island Press.

— (2005a) *Ecosystems and Human Well-Being: Wetlands and Water – Synthesis*, Washington, DC: Island Press

— (2005b) 'Living beyond our means: natural assets and human well-being', Statement of the MA board, www.millenniumassessment.org/documents/document.429.aspx.pdf.

Milne, R. and T. A. Brown (1997) 'Carbon in the vegetation and soils of Great Britain', *Journal of Environmental Management*, 49(4): 413–33.

Mittelman, J. (1998) 'Globalisation and environmental resistance politics', *Third World Quarterly*, 19(5): 847–72.

Moore, M. (2001) *Stupid White Men*, New York: HarperCollins.

Morris, J., T. M. Hess, D. J. G. Gowing, P. B. Leeds-Harrison, N. Bannister, R. M. N. Vivash and M. Wade (2005) 'A framework for integrating flood defence and biodiversity in washlands in England', *International Journal of River Basin Management*, 3(2): 1–11.

Moussavi-Nejad, E. (2003) 'Censuses of pastoral nomads and some general remarks about the census of nomadic tribes of Iran in 1998', *Nomadic Peoples*, 7(2): 24–35.

Murtough, G., B. Aretino and M. Anna (2002) 'Creating markets for ecosystem services', Productivity Commission Working Paper no. 1709, Australia, 17 June, ssrn.com/abstract=322380.

Naess, A. (1989) *Ecology, Community and Lifestyle*, Cambridge: Cambridge University Press.

Nam-choon, K. (2005) 'Ecological restoration and revegetation works in Korea', *Landscape and Ecological Engineering*, 1: 77–83.

Norgaard, R. B. (2010) 'Ecosystem services: from eye-opening metaphor to complexity blinder', *Ecological Economics*, 69: 1219–27.

Odum, W. (1982) 'Environmental degradation and the tyranny of small decisions', *BioScience*, 32(9): 728–9.

OECD (2010) *Paying for Biodiversity: Enhancing the Cost-Effectiveness of Payments for Ecosystem Services*, Paris: OECD Publishing.

Olsson, P., C. Folke, and T. P. Hughes, (2008) 'Navigating the transition to ecosystem-based management of the Great Barrier Reef, Australia', *Proceedings of the*

National Academy of Sciences USA, 105: 9489–94.

O'Neill, D. (2007) 'The total external environmental costs and benefits of agriculture in the UK', Environment Agency research report, www.environmentagency.gov. uk/static/documents/Research/costs_benefitapr07_1749472.pdf.

Opdam, P. and D. Wascher (2004) 'Climate change meets habitat fragmentation: linking landscape and biogeographical scale levels in research and conservation', *Biological Conservation*, 117(3): 285–97.

Ostrom, E. (1990) *Governing the Commons: The evolution of institutions for collective action*, Cambridge: Cambridge University Press.

Ostrom, E. and N. Dolšak (2003) *The Commons in the New Millennium: Challenges and Adaptations*, Cambridge, MA: MIT Press.

Ostrom, E., J. Burger, C. B. Field, R. B. Norgaard and D. Policansky (1999) 'Revisiting the commons: local lessons, global challenges', *Science*, 284: 278–82.

PAA (2010) *United Utilities SCaMP Sustainable Catchment Management Programme: Monitoring Progress Report Year 4*, Buxton: Penny Anderson Associates Ltd, 1 September, www.unitedutilities. com/scamp.htm.

Pagiola, S. (2008) 'Payments for environmental services in Costa Rica', *Ecological Economics*, 65(4): 712–24.

Pagiola, S., J. Bishop and N. Landell-Mills (eds) (2002) *Selling Forest Environmental Services: Market-based Mechanisms for Conservation and Development*, London: Earthscan.

Parker, D. J. (ed.) (2000) *Floods*, London: Routledge.

Pearce, D., A. Markandya and E. Barbier (1995) *Blueprint for a Green Economy*, London: Earthscan.

Perrot-Maître, D. (2006) 'The Vittel payments for ecosystem services: a "perfect" PES case?', Project Paper 3, London: International Institute for Environment and Development.

Petts, G., J. Heathcote and D. Martin (2001) *Urban Rivers: Our Inheritance and Future*, London: IWA Publishing.

Pitt, M. (2007) 'The Pitt Review: lessons learned from the 2007 floods', London: Cabinet Office, www.cabinetoffice.gov.uk/thepittreview/final_report.aspx.

Plunket Greene, H. (2007) *Where the Bright Waters Meet*, Ellesmere: Medlar Press.

Porritt, J. (2000) *Playing Safe: Science and the environment*, London: Thames and Hudson.

— (2005) *Capitalism as if the World Matters*, London: Earthscan.

Posey, D. A. (ed.) (2000) *Cultural and Spiritual Values of Biodiversity*, Rugby: ITDG Publishing.

Posner, R. A. (2009) *A Failure of Capitalism: The Crisis of '08 and the Descent into Depression*, Harvard, MA: Harvard University Press.

Posthumus, H., J. R. Rouquette, J. Morris, D. J. G. Gowing and T. M. Hess (2010) 'A framework for the assessment of ecosystem goods and services; a case study on lowland floodplains in England', *Ecological Economics*, 69(7): 1510–23.

Pretty, J., M. Griffin, J. Peacock, R. Hine, M. Sellens and N. South (2005) 'A countryside for health and wellbeing: the physical and mental health benefits of green exercise – executive summary', *Countryside Recreation*, 13: 2–7.

Prigogine, I. (1997) *The End of Certainty. Time, Chaos, and the New Laws of Nature*, New York: Free Press.

Ramsar Convention (1971) *Convention on Wetlands of International Importance Especially as Waterfowl Habitat*, 2 February, Ramsar, Iran.

Rangarajan, M. (2001) *India's Wildlife History*, Delhi: Permanent Black.

Republic of South Africa (1998) *National Water Act*, Act no. 36 of 1998, *Government Gazette*, Pretoria.

Roberts, P., C. George and J. Ravetz (2009) *Environment and the City*, London: Routledge.

Rodgers, C. P., E. A. Straughton, A. J. L. Winchester and M. Pieraccini (2010) *Contested Common Land; environmental governance past and present*, London: Earthscan.

Roszak, T., M. E. Gomes and A. D. Kanner (1995) *Ecopsychology: Restoring the Earth, Healing the Mind*, San Francisco, CA: Sierra Book Club.

Royal Society (2009) *Reaping the Benefits: Science and the Sustainable Intensification of Global Agriculture*, London: Royal Society.

RSPB (1970) 'Farming and wildlife: a study in compromise', Proceedings of a conference held at Silsoe, Sandy, Bedfordshire.

Sanderson, J. and L. D. Harris (eds) (2000) *Landscape Ecology: A Top-down Approach*, Boca Raton, FL: Lewis Publishers.

Schiemer, F. (2000) 'Fish as indicators for the assessment of the ecological integrity of large rivers', *Hydrobiologia*, 422/3: 271–8.

Scrase, J. I. and W. R. Sheate (2005) 'Re-framing flood control in England and Wales', *Environmental Values*, 14: 113–37.

Sheail, J. (2002) *An Environmental History of Twentieth-century Britain*, Basingstoke: Palgrave Macmillan.

Slater, G. (1907) *The English Peasantry and the Enclosure of Common Fields*, London: Archibald Constable and Co.

Sommerville, M. M., J. P. G. Jones and E. J. Milner-Gulland (2009) 'A revised conceptual framework for payments for environmental services', *Ecology and Society*, 14(2): 34.

Soulsby, C. (n.d.) 'Managing river habitats for fisheries: a guide to best practice', Scottish Environment Protection Agency, Stirling, www.millenniumassessment.org/en/SGA.Safma.aspx.

Spangenberg, J. H. and J. Settele (2010) 'Precisely incorrect? Monetising the value of ecosystem services', *Ecol Complex*, 7(3): 327–37.

Spence, M. D. (1999) *Dispossessing the Wilderness: Indian removal and the making of national parks*, Oxford: Oxford University Press.

Stirling, A. (2008) '"Opening up" and "closing down": power, participation, and pluralism in the social appraisal of technology', *Science, Technology and Human Values*, 33(2): 262–94.

Stirling, A., M. Leach, L. Mehta, I. Scoones, A. Smith, S. Stagl and J. Thompson (2007) 'Empowering designs: towards more progressive appraisal of sustainability', STEPS Working Paper 3, STEPS Centre, University of Brighton.

Strang, V. (1997) *Uncommon Ground: Landscape, Values and the Environment*, Oxford: Berg.

Swingland, I. (ed.) (2002) *Capturing Carbon and Conserving Biodiversity: The Market Approach*, London: Earthscan.

Taylor, P. (1998) *An Ecological*

Approach to International Law: Responding to Challenges of Climate Change, London: Routledge.

TEEB (2010) 'The economics of ecosystems and biodiversity: report for business – executive summary', www.teebweb.org/LinkClick.aspx?fileticket=ubcryEoOUbw%3d&tabid=1021&language=en-US.

Thompson, E. P. (1991) *The Making of the English Working Class*, London: Penguin.

Time Magazine (1932) 'The Roosevelt week', *Time*, 11 July.

Troll, C. (1939) *Luftbildplan und ökologische Bodenforschung* [Aerial photography and ecological studies of the earth], Berlin: Zeitschrift der Gesellschaft für Erdkunde.

Turner, M. G. (1989) 'Landscape ecology: the effect of pattern on process', *Annual Review of Ecology and Systematics*, 20: 171–97.

Turner, R. K. (1999) 'The place of economic values in environmental valuation', in I. J. Bateman and K. G. Willis (eds), *Valuing Environmental Preferences*, Oxford: Oxford University Press.

Turner, R. K., S. Georgiou and B. Fisher (2008) *Valuing Ecosystem Services: The Case of Multifunctional Wetlands*, London: Earthscan.

Turton, A. R., H. J. Hattinh, G. A. Maree, D. J. Roux, M. Claassen and W. F. Strydom (eds) (2006) *Governance as the Trialogue: Government-society-science in Transition (Water Resources Development and Management)*, Springer.

Tusa, E. (2000) 'Evaluation of economic benefits of the Tamar 2000 Project, Phase II', Report prepared for Westcountry Rivers Trust by the Wetland Ecosystems Research Group (WERG), Royal Holloway Institute for Environmental Research.

Ulrich, R. S. (1984) 'View through a window may influence recovery from surgery', *Science*, 224: 420–21.

UN Food and Agriculture Organization (1996) *Rome Declaration on World Food Security*, Rome: FAO.

UNDP (United Nations Development Programme) (2004) *Human Development Report 2004: Cultural Liberty in Today's Diverse World*, New York: UNDP.

UNEP (2005) *Creating Pro-poor Markets for Ecosystem Services*, Agreed notes from a high-level brainstorming workshop, London, 10–12 October, UNEP Division of Environmental Conventions.

US Department of Agriculture (2009) 'Forest land conversion, ecosystem services, and economic issues for policy: a review', Forests on the Edge report, www.fs.fed.us/openspace/fote/pnw-gtr797.pdf.

Van Koppen, B., M. Giordano and J. Butterworth (2008) 'Community-based water law and water resource management reform in developing countries', Comprehensive Assessment of Water Management in Agriculture Series 5, Wallingford: CABI.

Walker, G., J. Fairburn, G. Smith and G. Mitchell (2003) 'Environmental quality and social deprivation', R&D Technical Report, Bristol: Environment Agency.

Wanless, D. (2002) 'Securing our future health: taking a long-term view', London: Department of Health.

Wardle, D. A. (2005) 'Effects of biodiversity on ecosystem functioning: a consensus of current knowledge', *Ecological Monographs*, 75(1): 3–35.

Waverley Committee (1954) 'Recommendations of the Waverley Committee. Report of the Departmental Committee on Coastal Flooding (Cmd 9165)', London: HMSO.

WCED (World Commission on Environment and Development) (1987) *Our Common Future*, Oxford: Oxford University Press.

Weizsäcker, E. von, A. B. Lovins and L. H. Lovins (1997) *Factor Four: Doubling Wealth, Halving Resource Use – a New Report to the Club of Rome*, London: Earthscan.

Wilson, E. O. (1990) *Sociobiology*, Cambridge, MA: Harvard University Press.

— (1999) *Consilience*, London: Abacus.

Winchester, A. J. L. (2008) 'Statute and local custom: village byelaws and the governance of common land in medieval and early-modern England', Paper for IASC Global Conference, Cheltenham, July.

Woods-Ballard, B., R. Kellagher et al. (2007) *The SUDS Manual*, CIRIA Report C697, London: Construction Industry Research and Information Association.

Wordie, J. R. (1983) 'The chronology of English enclosure 1500–1914', *Economic History Review Series 2*, 36(4): 483–505.

World Bank (2007) *State and Trends of the Carbon Market 2007*, Washington, DC: World Bank.

World Commission on Dams (2000) *Dams and Development: A New Framework for Better Decision-making*, London: Earthscan.

World Resources Institute (WRI), United Nations Development Programme (UNDP), United Nations Environment Programme (UNEP) and World Bank (2005) *World Resources 2005: The Wealth of the Poor – Managing Ecosystems to Fight Poverty*, Washington, DC: World Resources Institute.

World Water Council (2006) 'Ecosystem and ecohydrology approaches to integrated water resources management', FT2.38, World Water Forum, Mexico, www.worldwatercouncil.org/index.php?id=1236&L=3.

Wunder, S. (2005) 'Payments for environmental services: some nuts and bolts', CIFOR Occasional Paper no. 42, Center for International Forestry Research, Bogor, Indonesia.

Wünscher, T., S. Engel and S. Wunder (2008) 'Spatial targeting of payments for environmental services: a tool for boosting conservation benefits', *Ecological Economics*, 65: 822–33.

WWF (2004) *Living Planet Report 2004*, Godalming: WWF, www.panda.org/downloads/general/lpr2004.pdf.

Yin, H. and C. Li (2001) 'Human impact on floods and flood disasters on the Yangtze river', *Geomorphology*, 41: 105–9.

Zeidan, D. (1995) *The Middle East Handbook: Information on States and People Groups in the Middle East*, vol. 3: *A–Z ME People Groups and Their Distribution*, OM-IRC, www.angelfire.com/az/rescon/MEHBKNMD.html.

Zimmerer, K. S. (1994) 'Human geography and the "new ecology": the prospect of promise and integration', *Annals of the Association of American Geography*, 84: 108–25.

Index